Introduction to Combustion Science

Introduction to Combustion Science

Edited by
Randall Fletcher

WILLFORD PRESS
www.willfordpress.com

Published by Willford Press,
118-35 Queens Blvd., Suite 400,
Forest Hills, NY 11375, USA

ISBN: 978-1-68285-484-6

Cataloging-in-Publication Data

Introduction to combustion science / edited by Randall Fletcher.
    p. cm.
Includes bibliographical references and index.
ISBN 978-1-68285-484-6
1. Combustion engineering. 2. Heat engineering. I. Fletcher, Randall.
TJ254.5 .I58 2018
621.402 3--dc23

For information on all Willford Press publications
visit our website at www.willfordpress.com

WILLFORD PRESS

Contents

Preface

Any reaction between a reductant and an oxidant, which leads to the formation of smoke, is called an exothermic redox chemical reaction. When this reaction takes place in high temperature it is termed as combustion. The different types of combustion are rapid, complete, micro-gravity, spontaneous, smoldering, incomplete, micro-combustion, etc. This book outlines the processes and applications of combustion in detail. The topics included in it are of utmost significance and bound to provide incredible insights to readers. Coherent flow of topics, student-friendly language and extensive use of examples make this text an invaluable source of knowledge.

A foreword of all Chapters of the book is provided below:

Chapter 1 - Combustion is an exothermic controlled reaction, which takes place between a fuel and an oxidizing agent. The mechanism of combustion can be explained with a combustion triangle. The materials that produce chemical or nuclear energy in the form of fire can be referred to as fuel. This chapter will provide an integrated understanding of combustion and fuel; **Chapter 2 -** Thermodynamics is a sub-field of physics that deals with heat and temperature. The related elements of thermodynamics are guided by the four laws of thermodynamics. It can be classified into two parts: intensive and extensive. While the former can be understood in the context of pressure, temperature and spectral entropy, the latter's properties include mass, volume and enthalpy. Combustion is best understood in confluence with the major topics listed in the following chapter; **Chapter 3 -** The process of combustion can be classified into thermodynamics, fluid mechanics, heat and mass transfer, and chemical kinetics. The main governing laws of combustion are Fick's law of diffusion, Newton's law of viscosity and Fourier's law of conduction. The aspects elucidated in this chapter are of vital importance, and provide a better understanding of transport phenomena in combustion; **Chapter 4 -** Many varied reactions take place in combustion. Chemical reaction can be categories in two ways –based on physical state of species and on the basis of reaction rate. The former includes homogenous reaction and heterogeneous reaction, and the latter into explosive and non-explosive. The reaction is dependent on the concentration of species, pressure and temperature. This chapter is an overview of the subject matter incorporating all the major aspects of combustion; **Chapter 5 -** A pre-mixed flame is formed during combustion of mixture of fuel and an oxidizer. It is necessary that the elements are mixed before they are made to combust. A pre-mixed flame can be characterized by its burning velocity. The topics discussed in the chapter are of great importance to broaden the existing knowledge on combustion; **Chapter 6 -** A diffusion flame can be formed when the oxidizer fuses with the fuel through the method of diffusion. Diffusion flame produces more soot and burns slower in comparison to pre-mixed flame. A few instances of diffusion flames are forest fire, solid fuel combustion, liquid fuel combustion, and candle flame. This chapter elucidates the principles of diffusion in combustion.

I would like to thank the entire editorial team who made sincere efforts for this book and my family who supported me in my efforts of working on this book. I take this opportunity to thank all those who have been a guiding force throughout my life.

Editor

An Introduction of Combustion and Fuel

Combustion is an exothermic controlled reaction, which takes place between a fuel and an oxidizing agent. The mechanism of combustion can be explained with a combustion triangle. The materials that produce chemical or nuclear energy in the form of fire can be referred to as fuel. This chapter will provide an integrated understanding of combustion and fuel.

Combustion

The flames caused as a result of a fuel undergoing combustion (burning)

Combustion or burning is a high-temperature exothermic redox chemical reaction between a fuel (the reductant) and an oxidant, usually atmospheric oxygen, that produces oxidized, often gaseous products, in a mixture termed as smoke. Combustion in a fire produces a flame, and the heat produced can make combustion self-sustaining. Combustion is often a complicated sequence of elementary radical reactions. Solid fuels, such as wood, first undergo endothermic pyrolysis to produce gaseous fuels whose combustion then supplies the heat required to produce more of them. Combustion is often hot enough that light in the form of either glowing or a flame is produced. A simple example can be seen in the combustion of hydrogen and oxygen into water vapor, a reaction commonly used to fuel rocket engines. This reaction releases 242 kJ/mol of heat and reduces the enthalpy accordingly (at constant temperature and pressure):

Air pollution abatement equipment provides combustion control for industrial processes.

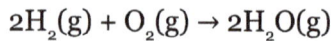

$$2H_2(g) + O_2(g) \rightarrow 2H_2O(g)$$

Combustion of an organic fuel in air is always exothermic because the double bond in O_2 is much weaker than other double bonds or pairs of single bonds, and therefore the formation of the stronger bonds in the combustion products CO_2 and H_2O results in the release of energy. The bond energies in the fuel play only a minor role, since they are similar to those in the combustion products; e.g., the sum of the bond energies of CH_4 is nearly the same as that of CO_2. The heat of combustion is approximately -418 kJ per mole of O_2 used up in the combustion reaction, and can be estimated from the elemental composition of the fuel.

Uncatalyzed combustion in air requires fairly high temperatures. Complete combustion is stoichiometric with respect to the fuel, where there is no remaining fuel, and ideally, no remaining oxidant. Thermodynamically, the chemical equilibrium of combustion in air is overwhelmingly on the side of the products. However, complete combustion is almost impossible to achieve, since the chemical equilibrium is not necessarily reached, or may contain unburnt products such as carbon monoxide, hydrogen and even carbon (soot or ash). Thus, the produced smoke is usually toxic and contains unburned or partially oxidized products. Any combustion at high temperatures in atmospheric air, which is 78 percent nitrogen, will also create small amounts of several nitrogen oxides, commonly referred to as NO_x, since the combustion of nitrogen is thermodynamically favored at high, but not low temperatures. Since combustion is rarely clean, flue gas cleaning or catalytic converters may be required by law.

Fires occur naturally, ignited by lightning strikes or by volcanic products. Combustion (fire) was the first controlled chemical reaction discovered by humans, in the form of campfires and bonfires, and continues to be the main method to produce energy for humanity. Usually, the fuel is carbon, hydrocarbons or more complicated mixtures such as wood that contains partially oxidized hydrocarbons. The thermal energy produced from combustion of either fossil fuels such as coal or oil, or from renewable fuels such as firewood, is harvested for diverse uses such as cooking, production of electricity or industrial or domestic heating. Combustion is also currently the only reaction used to power rockets. Combustion is also used to destroy (incinerate) waste, both nonhazardous and hazardous.

Oxidants for combustion have high oxidation potential and include atmospheric or pure oxygen, chlorine, fluorine, chlorine trifluoride, nitrous oxide and nitric acid. For instance, hydrogen burns in chlorine to form hydrogen chloride with the liberation of heat and light characteristic of combustion. Although usually not catalyzed, combustion can be catalyzed by platinum or vanadium, as in the contact process.

Types

Complete

$$CH_4 \; + \; 2O_2 \; \longrightarrow \; CO_2 \; + \; 2H_2O$$

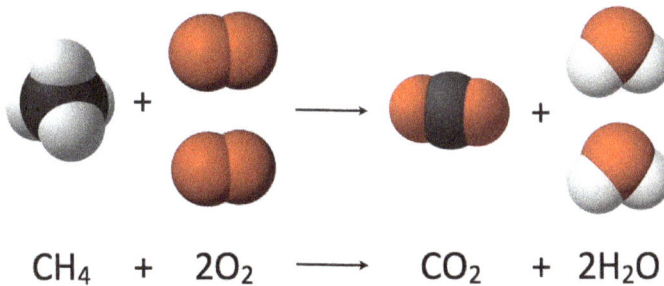

The combustion of methane, a hydrocarbon.

In complete combustion, the reactant burns in oxygen, producing a limited number of products. When a hydrocarbon burns in oxygen, the reaction will primarily yield carbon dioxide and water. When elements are burned, the products are primarily the most common oxides. Carbon will yield carbon dioxide, sulfur will yield sulfur dioxide, and iron will yield iron(III) oxide. Nitrogen is not considered to be a combustible substance when oxygen is the oxidant, but small amounts of various nitrogen oxides (commonly designated NO_x species) form when air is the oxidant.

Combustion is not necessarily favorable to the maximum degree of oxidation, and it can be temperature-dependent. For example, sulfur trioxide is not produced quantitatively by the combustion of sulfur. NOx species appear in significant amounts above about 2,800°F (1,540°C), and more is produced at higher temperatures. The amount of NOx is also a function of oxygen excess.

In most industrial applications and in fires, air is the source of oxygen (O_2). In air, each mole of oxygen is mixed with approximately 3.71 mol of nitrogen. Nitrogen does not take part in combustion, but at high temperatures some nitrogen will be converted to NO_x (mostly NO, with much smaller amounts of NO_2). On the other hand, when there is insufficient oxygen to completely combust the fuel, some fuel carbon is converted to carbon monoxide and some of the hydrogen remains unreacted. A more complete set of equations for the combustion of a hydrocarbon in air therefore requires an additional calculation for the distribution of oxygen between the carbon and hydrogen in the fuel.

The amount of air required for complete combustion to take place is known as theoretical air. However, in practice the air used is 2-3x that of theoretical air.

Incomplete

Incomplete combustion will occur when there is not enough oxygen to allow the fuel to react completely to produce carbon dioxide and water. It also happens when the combustion is quenched by a heat sink, such as a solid surface or flame trap.

For most fuels, such as diesel oil, coal or wood, pyrolysis occurs before combustion. In incomplete combustion, products of pyrolysis remain unburnt and contaminate the smoke with noxious particulate matter and gases. Partially oxidized compounds are also a concern; partial oxidation of ethanol can produce harmful acetaldehyde, and carbon can produce toxic carbon monoxide.

The quality of combustion can be improved by the designs of combustion devices, such as burners and internal combustion engines. Further improvements are achievable by catalytic after-burning devices (such as catalytic converters) or by the simple partial return of the exhaust gases into the combustion process. Such devices are required by environmental legislation for cars in most countries, and may be necessary to enable large combustion devices, such as thermal power stations, to reach legal emission standards.

The degree of combustion can be measured and analyzed with test equipment. HVAC contractors, firemen and engineers use combustion analyzers to test the efficiency of a burner during the combustion process. In addition, the efficiency of an internal combustion engine can be measured in this way, and some U.S. states and local municipalities use combustion analysis to define and rate the efficiency of vehicles on the road today.

Smouldering

Smouldering is the slow, low-temperature, flameless form of combustion, sustained by the heat evolved when oxygen directly attacks the surface of a condensed-phase fuel. It is a typically incomplete combustion reaction. Solid materials that can sustain a smouldering reaction include coal, cellulose, wood, cotton, tobacco, peat, duff, humus, synthetic foams, charring polymers (including polyurethane foam) and dust. Common examples of smouldering phenomena are the initiation of residential fires on upholstered furniture by weak heat sources (e.g., a cigarette, a short-circuited wire) and the persistent combustion of biomass behind the flaming fronts of wildfires.

Rapid

Rapid combustion is a form of combustion, otherwise known as a fire, in which large amounts of heat and light energy are released, which often results in a flame. This is used in a form of machinery such as internal combustion engines and in thermobaric weapons. Such a combustion is frequently called an explosion, though for an internal combustion engine this is inaccurate. An internal combustion engine nominally operates on a controlled rapid burn. When the fuel-air mixture in an internal combustion engine explodes, that is known as detonation.

Spontaneous

Spontaneous combustion is a type of combustion which occurs by self heating (increase in temperature due to exothermic internal reactions), followed by thermal runaway (self heating which rapidly accelerates to high temperatures) and finally, ignition. For example, phosphorus self-ignites at room temperature without the application of heat.

Turbulent

Combustion resulting in a turbulent flame is the most used for industrial application (e.g. gas turbines, gasoline engines, etc.) because the turbulence helps the mixing process between the fuel and oxidizer.

Micro-gravity

Colourized gray-scale composite image of the individual frames
from a video of a backlit fuel droplet burning in microgravity.

The term 'micro' gravity refers to a gravitational state that is 'low' (i.e., 'micro' in the sense of 'small' and not necessarily a millionth of Earth's normal gravity) such that the influence of buoyancy on physical processes may be considered small relative to other flow processes that would be present at normal gravity. In such an environment, the thermal and flow transport dynamics can behave quite differently than in normal gravity conditions (e.g., a candle's flame takes the shape of a sphere.). Microgravity combustion research contributes to the understanding of a wide variety of aspects that are relevant to both the environment of a spacecraft (e.g., fire dynamics relevant to crew safety on the International Space Station) and terrestrial (Earth-based) conditions (e.g., droplet combustion dynamics to assist developing new fuel blends for improved combustion, materials fabrication processes, thermal management of electronic systems, multiphase flow boiling dynamics, and many others).

Micro-combustion

Combustion processes which happen in very small volumes are considered micro-combustion. The high surface-to-volume ratio increases specific heat loss. Quenching distance plays a vital role in stabilizing the flame in such combustion chambers.

Chemical Equations

Stoichiometric Combustion of a Hydrocarbon in Oxygen

Generally, the chemical equation for stoichiometric combustion of a hydrocarbon in oxygen is:

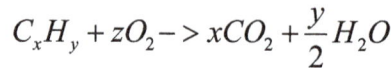

$$C_xH_y + zO_2 -> xCO_2 + \frac{y}{2}H_2O$$

where $z = x + y/4$.

For example, the stoichiometric burning of propane in oxygen is:

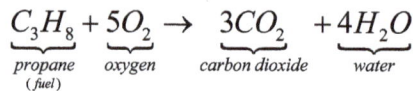

$$\underbrace{C_3H_8}_{\substack{propane \\ (fuel)}} + \underbrace{5O_2}_{oxygen} \rightarrow \underbrace{3CO_2}_{carbon\ dioxide} + \underbrace{4H_2O}_{water}$$

Stoichiometric Combustion of a Hydrocarbon in Air

If the stoichiometric combustion takes place using air as the oxygen source, the nitrogen present in the air (Atmosphere of Earth) can be added to the equation (although it does not react) to show the stoichiometric composition of the fuel in air and the composition of the resultant flue gas. Note that treating all non-oxygen components in air as nitrogen gives a 'nitrogen' to oxygen ration of 3.77, i.e. (100% - O_2%) / O_2% where O_2% is 20.95% vol:

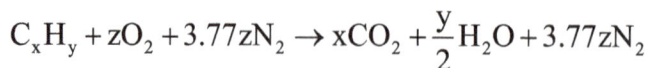

$$C_xH_y + zO_2 + 3.77zN_2 \rightarrow xCO_2 + \frac{y}{2}H_2O + 3.77zN_2$$

where $z = x + \frac{1}{4}y$.

For example, the stoichiometric combustion of propane (C3H8) in air is:

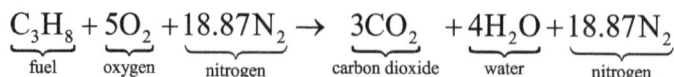

$$\underbrace{C_3H_8}_{fuel} + \underbrace{5O_2}_{oxygen} + \underbrace{18.87N_2}_{nitrogen} \rightarrow \underbrace{3CO_2}_{carbon\ dioxide} + \underbrace{4H_2O}_{water} + \underbrace{18.87N_2}_{nitrogen}$$

The stoichiometric composition of propane in air is 1 / (1 + 5 + 18.87) = 4.02% vol

Trace Combustion Products

Various other substances begin to appear in significant amounts in combustion products when the flame temperature is above about 1600 K. When excess air is used, nitrogen may oxidize to NO and, to a much lesser extent, to NO_2. CO forms by disproportionation of CO_2, and H_2 and OH form by disproportionation of H_2O.

For example, when 1 mol of propane is burned with 28.6 mol of air (120% of the stoichiometric amount), the combustion products contain 3.3% O_2. At 1400 K, the equilibrium combustion products contain 0.03% NO and 0.002% OH. At 1800 K, the combustion products contain 0.17% NO, 0.05% OH, 0.01% CO, and 0.004% H_2.

Diesel engines are run with an excess of oxygen to combust small particles that tend to form with only a stoichiometric amount of oxygen, necessarily producing nitrogen oxide emissions. Both the United States and European Union enforce limits to vehicle nitrogen oxide emissions, which necessitate the use of special catalytic converters or treatment of the exhaust with urea.

Incomplete Combustion of a Hydrocarbon in Oxygen

The incomplete (partial) combustion of a hydrocarbon with oxygen produces a gas mixture containing mainly CO_2, CO, H_2O, and H_2. Such gas mixtures are commonly prepared for use as protective atmospheres for the heat-treatment of metals and for gas carburizing. The general reaction equation for incomplete combustion of one mole of a hydrocarbon in oxygen is:

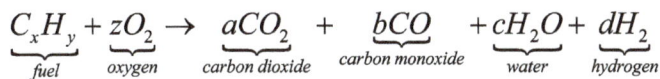

$$\underbrace{C_xH_y}_{fuel} + \underbrace{zO_2}_{oxygen} \rightarrow \underbrace{aCO_2}_{carbon\ dioxide} + \underbrace{bCO}_{carbon\ monoxide} + \underbrace{cH_2O}_{water} + \underbrace{dH_2}_{hydrogen}$$

When z falls below roughly 50% of the stoichiometric value, CH4 can become an important combustion product; when z falls below roughly 35% of the stoichiometric value, elemental carbon may become stable.

The products of incomplete combustion can be calculated with the aid of a material balance, together with the assumption that the combustion products reach equilibrium. For example, in the combustion of one mole of propane (C_3H_8) with four moles of O_2, seven moles of combustion gas are formed, and z is 80% of the stoichiometric value. The three elemental balance equations are:

- Carbon: $a + b = 3$

- Hydrogen: $2c + 2d = 8$

- Oxygen: $2a + b + c = 8$

These three equations are insufficient in themselves to calculate the combustion gas composition. However, at the equilibrium position, the water-gas shift reaction gives another equation:

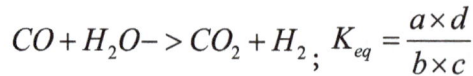

$$CO + H_2O -> CO_2 + H_2; \quad K_{eq} = \frac{a \times d}{b \times c}$$

For example, at 1200 K the value of K_{eq} is 0.728. Solving, the combustion gas consists of 42.4% H_2O, 29.0% CO_2, 14.7% H_2, and 13.9% CO. Carbon becomes a stable phase at 1200 K and 1 atm pressure when z is less than 30% of the stoichiometric value, at which point the combustion products contain more than 98% H_2 and CO and about 0.5% CH_4.

Fuels

Substances or materials which undergo combustion are called fuels. The most common examples are natural gas, propane, kerosene, diesel, petrol, charcoal, coal, wood, etc.

Liquid Fuels

Combustion of a liquid fuel in an oxidizing atmosphere actually happens in the gas phase. It is the vapor that burns, not the liquid. Therefore, a liquid will normally catch fire only above a certain temperature: its flash point. The flash point of a liquid fuel is the lowest temperature at which it can form an ignitable mix with air. It is the minimum temperature at which there is enough evaporated fuel in the air to start combustion.

Gaseous Fuels

Combustion of gaseous fuels may occur through one of four distinctive types of burning: diffusion flame, premixed flame, autoignitive reaction front, or as a detonation. The type of burning that actually occurs depends on the degree to which the fuel and oxidizer are mixed prior to heating: for example, a diffusion flame is formed if the fuel and oxidizer are separated initially, whereas a premixed flame is formed otherwise. Similarly, the type of burning also depends on the pressure: a detonation, for example, is an autoignitive reaction front coupled to a strong shock wave giving it its characteristic high-pressure peak and high detonation velocity.

Solid Fuels

The act of combustion consists of three relatively distinct but overlapping phases:

- Preheating phase, when the unburned fuel is heated up to its flash point and then fire point. Flammable gases start being evolved in a process similar to dry distillation.

- Distillation phase or gaseous phase, when the mix of evolved flammable gases with oxygen is ignited. Energy is produced in the form of heat and light. Flames are often visible. Heat transfer from the combustion to the solid maintains the evolution of flammable vapours.

- Charcoal phase or solid phase, when the output of flammable gases from the material is too low for persistent presence of flame and the charred fuel does not burn rapidly and just glows and later only smoulders.

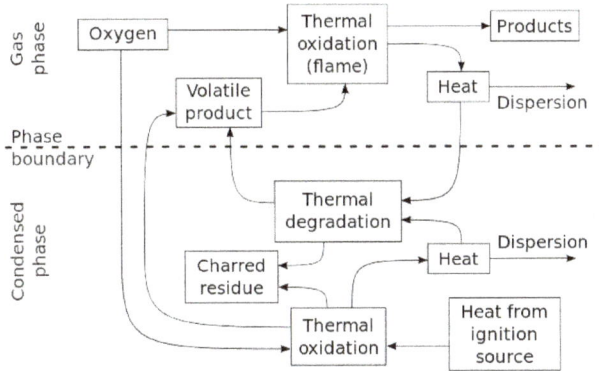

A general scheme of polymer combustion

Combustion Management

Efficient process heating requires recovery of the largest possible part of a fuel's heat of combustion into the material being processed. There are many avenues of loss in the operation of a heating process. Typically, the dominant loss is sensible heat leaving with the offgas (i.e., the flue gas). The temperature and quantity of offgas indicates its heat content (enthalpy), so keeping its quantity low minimizes heat loss.

In a perfect furnace, the combustion air flow would be matched to the fuel flow to give each fuel molecule the exact amount of oxygen needed to cause complete combustion. However, in the real world, combustion does not proceed in a perfect manner. Unburned fuel (usually CO and H_2) discharged from the system represents a heating value loss (as well as a safety hazard). Since combustibles are undesirable in the offgas, while the presence of unreacted oxygen there presents minimal safety and environmental concerns, the first principle of combustion management is to provide more oxygen than is theoretically needed to ensure that all the fuel burns. For methane (CH_4) combustion, for example, slightly more than two molecules of oxygen are required.

The second principle of combustion management, however, is to not use too much oxygen. The correct amount of oxygen requires three types of measurement: first, active control of air and fuel flow; second, offgas oxygen measurement; and third, measurement of offgas combustibles. For each heating process there exists an optimum condition of minimal offgas heat loss with acceptable levels of combustibles concentration. Minimizing excess oxygen pays an additional benefit: for a given offgas temperature, the NOx level is lowest when excess oxygen is kept lowest.

Adherence to these two principles is furthered by making material and heat balances on the combustion process. The material balance directly relates the air/fuel ratio to the percentage of O_2 in the combustion gas. The heat balance relates the heat available for the charge to the overall net heat produced by fuel combustion. Additional material and heat balances can be made to quantify the thermal advantage from preheating the combustion air, or enriching it in oxygen.

Reaction Mechanism

Combustion in oxygen is a chain reaction in which many distinct radical intermediates participate. The high energy required for initiation is explained by the unusual structure of the dioxygen molecule. The lowest-energy configuration of the dioxygen molecule is a stable, relatively unreactive diradical in a triplet spin state. Bonding can be described with three bonding electron pairs and two antibonding electrons, whose spins are aligned, such that the molecule has nonzero total angular momentum. Most fuels, on the other hand, are in a singlet state, with paired spins and zero total angular momentum. Interaction between the two is quantum mechanically a "forbidden transition", i.e. possible with a very low probability. To initiate combustion, energy is required to force dioxygen into a spin-paired state, or singlet oxygen. This intermediate is extremely reactive. The energy is supplied as heat, and the reaction then produces additional heat, which allows it to continue.

Combustion of hydrocarbons is thought to be initiated by hydrogen atom abstraction (not proton abstraction) from the fuel to oxygen, to give a hydroperoxide radical (HOO). This reacts further to give hydroperoxides, which break up to give hydroxyl radicals. There are a great variety of these processes that produce fuel radicals and oxidizing radicals. Oxidizing species include singlet oxygen, hydroxyl, monatomic oxygen, and hydroperoxyl. Such intermediates are short-lived and cannot be isolated. However, non-radical intermediates are stable and are produced in incomplete combustion. An example is acetaldehyde produced in the combustion of ethanol. An intermediate in the combustion of carbon and hydrocarbons, carbon monoxide, is of special importance because it is a poisonous gas, but also economically useful for the production of syngas.

Solid and heavy liquid fuels also undergo a great number of pyrolysis reactions that give more easily oxidized, gaseous fuels. These reactions are endothermic and require constant energy input from the ongoing combustion reactions. A lack of oxygen or other poorly designed conditions result in these noxious and carcinogenic pyrolysis products being emitted as thick, black smoke.

The rate of combustion is the amount of a material that undergoes combustion over a period of time. It can be expressed in grams per second (g/s) or kilograms per second (kg/s).

Detailed descriptions of combustion processes, from the chemical kinetics perspective, requires the formulation of large and intricate webs of elementary reactions. For instance, combustion of hydrocarbon fuels typically involve hundreds of chemical species reacting according to thousands of reactions

Inclusion of such mechanisms within computational flow solvers still represents a pretty challenging task mainly in two aspects. First, the number of degrees of freedom (proportional to the number of chemical species) can be dramatically large; second the source term due to reactions introduces a disparate number of time scales which makes the whole dynamical system stiff. As a result, the direct numerical simulation of turbulent reactive flows with heavy fuels soon becomes intractable even for modern supercomputers.

Therefore, a plethora of methodologies has been devised for reducing the complexity of combustion mechanisms without renouncing to high detail level. Examples are provided by: the Relaxation Redistribution Method (RRM) The Intrinsic Low-Dimensional Manifold (ILDM) approach and further developments The invariant constrained equilibrium edge preimage curve method. A few variational approaches The Computational Singular perturbation (CSP) method and further developments. The Rate Controlled Constrained Equilibrium (RCCE) and Quasi Equilibrium Manifold (QEM) approach. The G-Scheme. The Method of Invariant Grids (MIG).

Temperature

Antoine Lavoisier conducting an experiment related combustion generated by amplified sun light.

Assuming perfect combustion conditions, such as complete combustion under adiabatic conditions (i.e., no heat loss or gain), the adiabatic combustion temperature can be determined. The formula that yields this temperature is based on the first law of thermodynamics and takes note of the fact that the heat of combustion is used entirely for heating the fuel, the combustion air or oxygen, and the combustion product gases (commonly referred to as the *flue gas*).

In the case of fossil fuels burnt in air, the combustion temperature depends on all of the following:

- the heating value;

- the stoichiometric air to fuel ratio λ;

- the specific heat capacity of fuel and air;

- the air and fuel inlet temperatures.

The adiabatic combustion temperature (also known as the *adiabatic flame tempera-ture*) increases for higher heating values and inlet air and fuel temperatures and for stoichiometric air ratios approaching one.

Most commonly, the adiabatic combustion temperatures for coals are around 2,200 °C (3,992 °F) (for inlet air and fuel at ambient temperatures and for $\lambda = 1.0$), around 2,150 °C (3,902 °F) for oil and 2,000 °C (3,632 °F) for natural gas.

In industrial fired heaters, power stationsteam generators, and large gas-fired tur-bines, the more common way of expressing the usage of more than the stoichiometric combustion air is *percent excess combustion air*. For example, excess combustion air of 15 percent means that 15 percent more than the required stoichiometric air is being used.

Instabilities

Combustion instabilities are typically violent pressure oscillations in a combustion chamber. These pressure oscillations can be as high as 180 dB, and long term expo-sure to these cyclic pressure and thermal loads reduces the life of engine components. In rockets, such as the F1 used in the Saturn V program, instabilities led to massive damage of the combustion chamber and surrounding components. This problem was solved by re-designing the fuel injector. In liquid jet engines the droplet size and distri-bution can be used to attenuate the instabilities. Combustion instabilities are a major concern in ground-based gas turbine engines because of NOx emissions. The tendency is to run lean, an equivalence ratio less than 1, to reduce the combustion temperature and thus reduce the NOx emissions; however, running the combustion lean makes it very susceptible to combustion instability.

The Rayleigh Criterion is the basis for analysis of thermoacoustic combustion instabil-ity and is evaluated using the Rayleigh Index over one cycle of instability

$$G(x) = \frac{1}{T}\int_T q'(x,t)p'(x,t)dt$$

where q' is the heat release rate perturbation and p' is the pressure fluctuation. When the heat release oscillations are in phase with the pressure oscillations, the Rayleigh Index is positive and the magnitude of the thermo acoustic instability is maximised. On the other hand, if the Rayleigh Index is negative, then thermoacoustic damping occurs. The Rayleigh Criterion implies that a thermoacoustic instability can be optimally controlled by having heat release oscillations 180 degrees out of phase with pressure oscillations at the same frequency. This minimizes the Rayleigh Index.

Combustion Triangle

Essential conditions for combustion to occur

1. Presence of fuel .

2. Presence of oxidizer (Not essentially oxygen).

3. They must be in right proportions.

4. The proportion will be dictated by flammability limit.

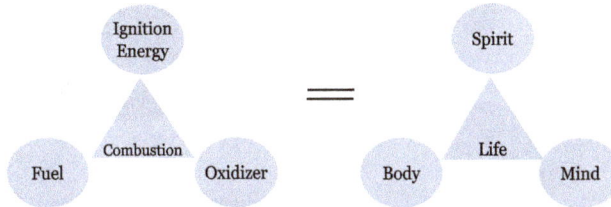

5. Ignition energy.

Industrial Process

- Thermal energy for process chemical plants, sugar industries, food processing industries are obtained through combustion.

- Iron, steel and other metals are produced from raw materials through combustion.

- Heat treatment and annealing of metals.

- Rotary kilns are used to produce Portlant cement.

Sugar Industry   Food Processing

Process Chemical Plant   Steel Plant

Transportation

- Surface transport vehicles are operated by reciprocating IC Engines.

- Gas turbine combustors are used widely in air and marine transportation sectors.

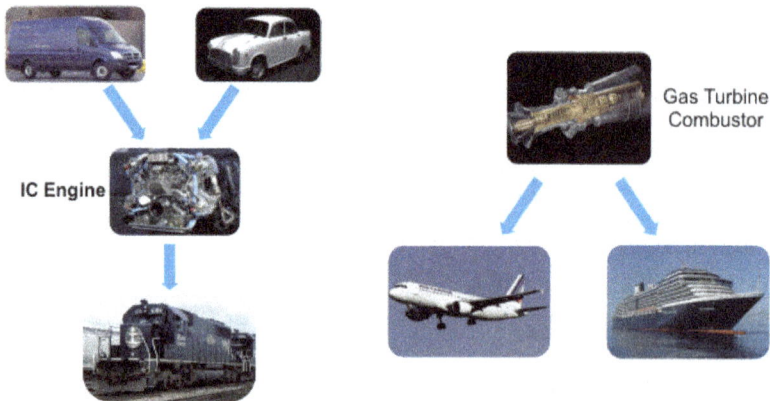

Power Generation

- Most of the thermal power plants are operated by burning coal.

- Recently gas turbine power plants are coming up.

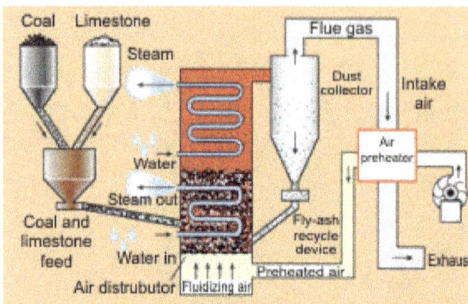

Fluidized Bed Power Plant   Coal Power Plant

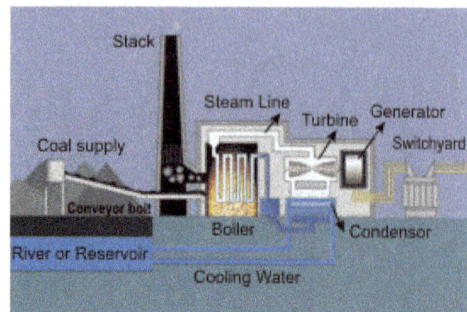

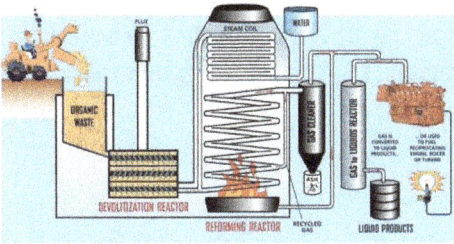

Biomass Based Power Plant

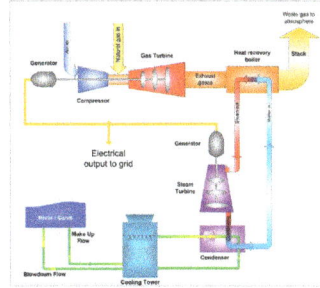

Gas Turbine Power Plant

Waste Disposal

- Combustion finds application in disposing waste materials.

- Incinerators are used to dispose domestic and industrial wastes.

- In modern hospitals, incinerators are used to dispose hospital wastes safely.

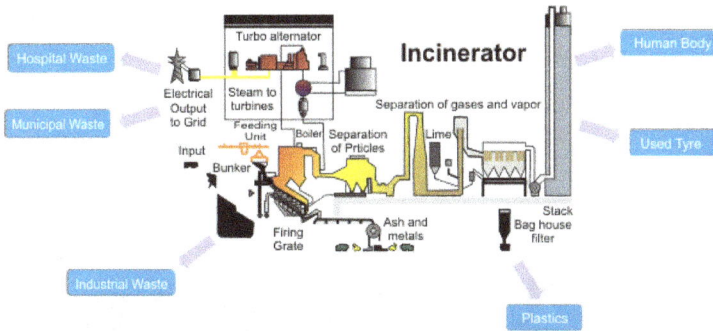

Fire

- Sometimes fire causes damage to life and property.

- Forest Fire: Damages natural resources and lives.

- Structural Fire: Damages property and human lives.

- Effective fire breakers should be designed and implemented to avoid fire hazard. By using proper construction materials, Fire hazard can be minimized. Marine life is very much affected by oil spill fire.

Environmental Pollution

- Combustion of any fuel produces certain amount of emissions such as smoke, ash, soot, and other harmful gases.

- Major pollution generated in combustion system are CO, CO_2, NO, NO_2, SO_2, ash, etc.

- These are due to incomplete combustion and can be minimized by increasing the residence time of fuel-oxidizer mixture in the combustor.

Transportation	7	79	60	15	6
Stationary Combustion Systems/Electricity	8	<1	<1	2	69
Industrial Process	23	8	32	13	25
Miscellaneous	62	13	8	70	<1

Fuel

Wood was one of the first fuels to be used by humans and is still the primary energy source in much of the world.

A fuel is any material that can be made to react with other substances so that it releases chemical or nuclear energy as heat or to be used for work. The concept was originally applied solely to those materials capable of releasing chemical energy but has since also been applied to other sources of heat energy such as nuclear energy (via nuclear fission and nuclear fusion).

The heat energy released by reactions of fuels is converted into mechanical energy via a heat engine. Other times the heat itself is valued for warmth, cooking, or industrial processes, as well as the illumination that comes with combustion. Fuels are also used in the cells of organisms in a process known as cellular respiration, where organic molecules are oxidized to release usable energy. Hydrocarbons and related oxygen-containing molecules are by far the most common source of fuel used by humans, but other substances, including radioactive metals, are also utilized.

Fuels are contrasted with other substances or devices storing potential energy, such as those that directly release electrical energy (such as batteries and capacitors) or mechanical energy (such as flywheels, springs, compressed air, or water in a reservoir).

History

The first known use of fuel was the combustion of wood or sticks by *Homo erectus* near 2,000,000 (two million) years ago. Throughout most of human history fuels derived from plants or animal fat were only used by humans. Charcoal, a wood derivative, has been used since at least 6,000 BCE for melting metals. It was only supplanted by coke, derived from coal, as European forests started to become depleted around the 18th century. Charcoal briquettes are now commonly used as a fuel for barbecue cooking.

Coal was first used as a fuel around 1000 BCE in China. With the energy in the form of chemical energy that could be released through combustion, but the concept development of the steam engine in the United Kingdom in 1769, coal came into more common use as a power source. Coal was later used to drive ships and locomotives. By the 19th century, gas extracted from coal was being used for street lighting in London. In the 20th and 21st centuries, the primary use of coal is to generate electricity, providing 40% of the world's electrical power supply in 2005.

Fossil fuels were rapidly adopted during the industrial revolution, because they were more concentrated and flexible than traditional energy sources, such as water power. They have become a pivotal part of our contemporary society, with most countries in the world burning fossil fuels in order to produce power.

Currently the trend has been towards renewable fuels, such as biofuels like alcohols.

Chemical

Chemical fuels are substances that release energy by reacting with substances around them, most notably by the process of combustion. Most of the chemical energy released in combustion was not stored in the chemical bonds of the fuel, but in the weak double bond of molecular oxygen.

Chemical fuels are divided in two ways. First, by their physical properties, as a solid, liquid or gas. Secondly, on the basis of their occurrence: *primary (natural fuel)* and *secondary (artificial fuel)*. Thus, a general classification of chemical fuels is:

General types of chemical fuels		
	Primary (natural)	**Secondary (artificial)**
Solid fuels	wood, coal, peat, dung, etc.	coke, charcoal
Liquid fuels	petroleum	diesel, gasoline, kerosene, LPG, coal tar, naphtha, ethanol
Gaseous fuels	natural gas	hydrogen, propane, methane, coal gas, water gas, blast furnace gas, coke oven gas, CNG

Solid Fuel

Coal is an important solid fuel

Solid fuel refers to various types of solid material that are used as fuel to produce energy and provide heating, usually released through combustion. Solid fuels include wood, charcoal, peat, coal, Hexamine fuel tablets, and pellets made from wood, corn, wheat, rye and other grains. Solid-fuel rocket technology also uses solid fuel. Solid fuels have been used by humanity for many years to create fire. Coal was the fuel source which enabled the industrial revolution, from firing furnaces, to running steam engines. Wood was also extensively used to run steam locomotives. Both peat and coal are still used in electricity generation today. The use of some solid fuels (e.g. coal) is restricted or prohibited in some urban areas, due to unsafe levels of toxic emissions. The use of other solid fuels such as wood is decreasing as heating technology and the availability of good quality fuel improves. In some areas, smokeless coal is often the only solid fuel used. In Ireland, peat briquettes are used as smokeless fuel. They are also used to start a coal fire.

Liquid Fuels

Liquid fuels are combustible or energy-generating molecules that can be harnessed to create mechanical energy, usually producing kinetic energy; they also must take the shape of their container. It is the fumes of liquid fuels that are flammable instead of the fluid.

A gasoline station

Most liquid fuels in widespread use are derived from the fossilized remains of dead plants and animals by exposure to heat and pressure in the Earth's crust. However, there are several types, such as hydrogen fuel (for automotive uses), ethanol, jet fuel and biodiesel which are all categorized as a liquid fuel. Emulsified fuels of oil-in-water such as orimulsion have been developed a way to make heavy oil fractions usable as liquid fuels. Many liquid fuels play a primary role in transportation and the economy.

Some common properties of liquid fuels are that they are easy to transport, and can be handled with relative ease. Also they are relatively easy to use for all engineering applications, and home use. Fuels like kerosene are rationed in some countries, for example available in government subsidized shops in India for home use.

Conventional diesel is similar to gasoline in that it is a mixture of aliphatic hydrocarbons extracted from petroleum. Kerosene is used in kerosene lamps and as a fuel for cooking, heating, and small engines. Natural gas, composed chiefly of methane, can only exist as a liquid at very low temperatures (regardless of pressure), which limits its direct use as a liquid fuel in most applications. LP gas is a mixture of propane and butane, both of which are easily compressible gases under standard atmospheric conditions. It offers many of the advantages of compressed natural gas (CNG), but is denser than air, does not burn as cleanly, and is much more easily compressed. Commonly used for cooking and space heating, LP gas and compressed propane are seeing increased use in motorized vehicles; propane is the third most commonly used motor fuel globally.

Gaseous Fuels

Fuel gas is any one of a number of fuels that under ordinary conditions are gaseous. Many fuel gases are composed of hydrocarbons (such as methane or propane), hydrogen, carbon monoxide, or mixtures thereof. Such gases are sources of potential heat energy or light energy that can be readily transmitted and distributed through pipes from the point of origin directly to the place of consumption. Fuel gas is contrasted with liquid fuels and from solid fuels, though some fuel gases are liquefied for storage or transport. While their gaseous nature can be advantageous, avoiding the difficulty of transporting solid fuel and the dangers of spillage inherent in liquid fuels, it can also be dangerous. It is possible for a fuel gas to be undetected and collect in certain areas, leading to the risk of a gas explosion. For this reason, odorizers are added to most fuel gases so that they may be detected by a distinct smell. The most common type of fuel gas in current use is natural gas.

A 20-pound (9.1 kg) propane cylinder

Biofuels

Biofuel can be broadly defined as solid, liquid, or gas fuel consisting of, or derived from biomass. Biomass can also be used directly for heating or power—known as *biomass fuel*. Biofuel can be produced from any carbon source that can be replenished rapidly e.g. plants. Many different plants and plant-derived materials are used for biofuel manufacture.

Perhaps the earliest fuel employed by humans is wood. Evidence shows controlled fire was used up to 1.5 million years ago at Swartkrans, South Africa. It is unknown which hominid species first used fire, as both *Australopithecus* and an early species of *Homo* were present at the sites. As a fuel, wood has remained in use up until the present day, although it has been superseded for many purposes by other sources. Wood has an energy density of 10–20 MJ/kg.

Recently biofuels have been developed for use in automotive transport (for example Bioethanol and Biodiesel), but there is widespread public debate about how carbon efficient these fuels are.

Fossil Fuels

Extraction of petroleum

Fossil fuels is hydrocarbons, primarily coal and petroleum (liquid petroleum or natural gas), formed from the fossilized remains of ancient plants and animals by exposure to high heat and pressure in the absence of oxygen in the Earth's crust over hundreds of millions of years. Commonly, the term fossil fuel also includes hydrocarbon-containing natural resources that are not derived entirely from biological sources, such as tar sands. These latter sources are properly known as *mineral fuels.*

Fossil fuels contain high percentages of carbon and include coal, petroleum, and natural gas. They range from volatile materials with low carbon:hydrogen ratios like methane, to liquid petroleum to nonvolatile materials composed of almost pure carbon, like anthracite coal. Methane can be found in hydrocarbon fields, alone, associated with oil, or in the form of methane clathrates. Fossil fuels formed from the fossilized remains of dead plants by exposure to heat and pressure in the Earth's crust over millions of years. This biogenic theory was first introduced by German scholar Georg Agricola in 1556 and later by Mikhail Lomonosov in the 18th century.

It was estimated by the Energy Information Administration that in 2007 primary sources of energy consisted of petroleum 36.0%, coal 27.4%, natural gas 23.0%, amounting to an 86.4% share for fossil fuels in primary energy consumption in the world. Non-fossil sources in 2006 included hydroelectric 6.3%, nuclear 8.5%, and others (geothermal, solar, tidal, wind, wood, waste) amounting to 0.9%. World energy consumption was growing about 2.3% per year.

Fossil fuels are non-renewable resources because they take millions of years to form, and reserves are being depleted much faster than new ones are being made. So we must conserve these fuels and use them judiciously. The production and use of fossil fuels raise environmental concerns. A global movement toward the generation of renewable energy is therefore under way to help meet increased energy needs. The burning of fossil fuels produces around 21.3 billion tonnes (21.3 gigatonnes) of carbon dioxide (CO_2) per year, but it is estimated that natural processes can only absorb about half of that amount, so there is a net increase of 10.65 billion tonnes of atmospheric carbon dioxide per year (one tonne of atmospheric carbon is equivalent to 44/12 or 3.7 tonnes of carbon dioxide). Carbon dioxide is one of the greenhouse gases that enhances radiative forcing and contributes to global warming, causing the average surface temperature of the Earth to rise in response, which the vast majority of climate scientists agree will cause major adverse effects. Fuels are a source of energy.

Energy

The amount of energy from different types of fuel depends on the stoichiometric ratio, the chemically correct air and fuel ratio to ensure complete combustion of fuel, and its Specific energy, the energy per unit mass.

Energy Capacities of Common Types of Fuel				
Fuel	Specific energy (MJ/kg)	AFR stoich.	FAR stoich.	Energy @ $\lambda=1$ (MJ/kg$_{(Air)}$)
Diesel	48	14.5 : 1	0.069 : 1	3.310
Ethanol	26.4	9 : 1	0.111 : 1	2.933
Gasoline	46.4	14.7 : 1	0.068 : 1	3.156
Hydrogen	142	34.3 : 1	0.029 : 1	4.140
Kerosene	46	15.6 : 1	0.064 : 1	2.949
LPG	46.4	17.2 : 1	0.058 : 1	2.698
Methanol	19.7	6.47 : 1	0.155 : 1	3.045
Nitromethane	11.63	1.7 : 1	0.588 : 1	6.841

1 MJ ≈ 0.28 kWh ≈ 0.37 HPh.

Nuclear

CANDU fuel bundles Two CANDU ("CANada Deuterium Uranium")
fuel bundles, each about 50 cm long and 10 cm in diameter

Nuclear fuel is any material that is consumed to derive nuclear energy. Technically speaking, All matter can be a nuclear fuel because any element under the right conditions will release nuclear energy, but the materials commonly referred to as nuclear fuels are those that will produce energy without being placed under extreme duress. Nuclear fuel is a material that can be 'burned' by nuclear fission or fusion to derive nuclear energy. *Nuclear fuel* can refer to the fuel itself, or to physical objects (for example bundles composed of fuel rods) composed of the fuel material, mixed with structural, neutron moderating, or neutron reflecting materials.

Most nuclear fuels contain heavy fissile elements that are capable of nuclear fission. When these fuels are struck by neutrons, they are in turn capable of emitting neutrons when they break apart. This makes possible a self-sustaining chain reaction that releases energy with a controlled rate in a nuclear reactor or with a very rapid uncontrolled rate in a nuclear weapon.

The most common fissile nuclear fuels are uranium-235 (^{235}U) and plutonium-239 (^{239}Pu). The actions of mining, refining, purifying, using, and ultimately disposing of nuclear fuel together make up the nuclear fuel cycle. Not all types of nuclear fuels create power from nuclear fission. Plutonium-238 and some other elements are used to produce small amounts of nuclear power by radioactive decay in radioisotope thermoelectric generators and other types of atomic batteries. Also, light nuclides such as tritium (^{3}H) can be used as fuel for nuclear fusion. Nuclear fuel has the highest energy density of all practical fuel sources.

Fission

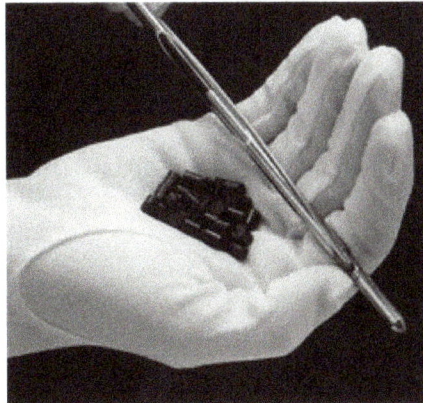

Nuclear fuel pellets are used to release nuclear energy

The most common type of nuclear fuel used by humans is heavy fissile elements that can be made to undergo nuclear fission chain reactions in a nuclear fission reactor; *nuclear fuel* can refer to the material or to physical objects (for example fuel bundles composed of fuel rods) composed of the fuel material, perhaps mixed with structural, neutron moderating, or neutron reflecting materials. The most common fissile nuclear fuels are U and Pu, and the actions of mining, refining, purifying, using, and ultimately disposing of these elements together make up the nuclear fuel cycle, which is important for its relevance to nuclear power generation and nuclear weapons.

Fusion

Fuels that produce energy by the process of nuclear fusion are currently not utilized by humans but are the main source of fuel for stars. Fusion fuels tend to be light elements such as hydrogen which will combine easily. Energy is required to start fusion by raising temperature so high all materials would turn into plasma, and allow nuclei to collide and stick together with each other before repelling due to electric charge. This process is called fusion and it can give out energy.

In stars that undergo nuclear fusion, fuel consists of atomic nuclei that can release energy by the absorption of a proton or neutron. In most stars the fuel is provided by hydrogen, which can combine to form helium through the proton-proton chain reaction or by the CNO cycle. When the hydrogen fuel is exhausted, nuclear fusion can continue with progressively heavier elements, although the net energy released is lower because of the smaller difference in nuclear binding energy. Once iron-56 or nickel-56 nuclei are produced, no further energy can be obtained by nuclear fusion as these have the highest nuclear binding energies. The elements then on use up energy instead of giving off energy when fused. Therefore, fusion stops and the star dies. In attempts by humans, fusion is only carried out with hydrogen (isotope of 2 and 3) to form helium-4 as this reaction gives out the most net energy. Electric confinement (ITER), inertial confinement(heating by laser) and heating by strong electric currents are the popular methods used.

World Trade

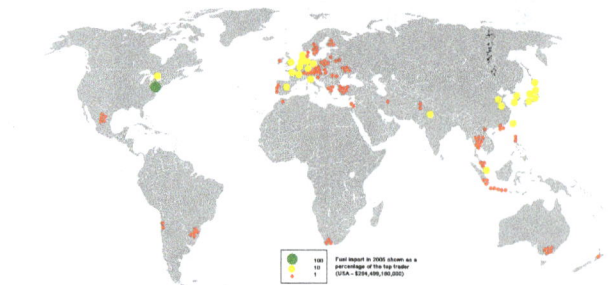

Fuel imports in 2005

The World Bank reported that the USA was the top fuel importer in 2005 followed by the EU and Japan.

Solid Fuels and Oxidizers

Solid Fuels

- Wood
- Coal
- Charcoal
- Soft Coke
- Biomass
- Animal dung

Constituents of Solid Fuel

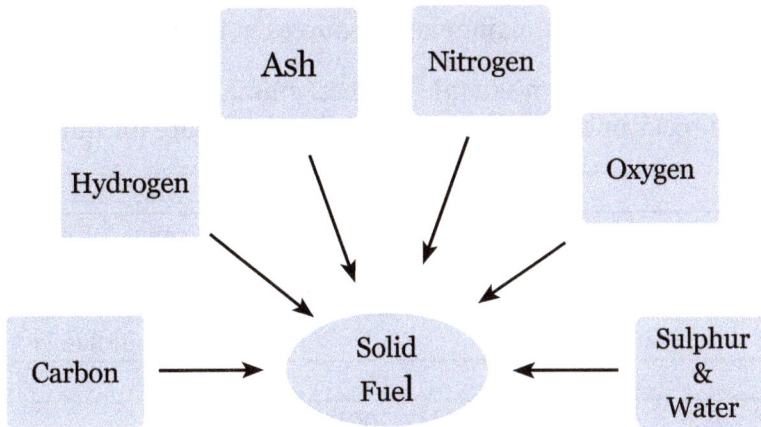

Types of Solid Fuels and Oxidizers

1	Biomass (Wood, Saw Dust, Rice Husk, Rice Straw, Wheat Straw, etc)	Air/O_2	Domestic Burner, Engine With Producer Gas
2	Coal, Coke, Charcoal	do	do
3	Special Fuels Nitrocellulose (NC), HTPB, CTPB	Nitroglycerine, Ammonium Perchlorate , Ammonium Nitrate, Nitrogen Tetraoxide	Solid Propellant Rocket, Hybrid Rocket

Oxygen, Water and Ash Content of Certain Solid Fuels

Moisture in Solid Fuel

1. Free

2. Bound water

Fuel moisture content will affect rate of combustion and overall efficiency.

Ash: The inorganic materials, which remain as residue even after complete combustion.

Ash content affects the performance of the combustion system.

Ash content causes fouling of the boilers.

Wood	40-45%	15-70%	0.1-1.0%
Peat	30-35%	70-90%	0.1-20%
Lignite coal	20-25%	20-30%	>5%
Bituminous coal	3-5%	10-5%	>5%
Anthracite coal	1-2%	2-4%	>5%

Liquid Fuels and Oxidizers

- Liquid fuel is one of the major energy sources in the transport sector.

- Crude oil is formed from organic sources, animals, vegetables – which are entrapped in rocks under high pressure and temperature for million years.

1	Gasoline	Air	S.I. Engine, Aircraft Piston Engine
2	HSD	Air	C.I. Engine
3	Furnace Oil	Air	Furnaces
4	Kerosene	Air	Aircraft, Gas Turbine, Engines Ramjet, Domestic Burner
5	Alcohols	Air	I.C. Engine
6	Hydrazine, UDMH, MMH, Liquid Hydrogen, Triethyl Amine	Liquid O_2, RFNA (Red Fuming Nitric Acid) N_2O_4	Liquid propellant rocket Engines
7	Hydrogen, Kerocene	Air	Ramjet/Scramjet

Refinery End-products of Typical Crude Oil

- Crude oil undergoes several process in the refinery.

- Generally separation of petroleum constituents occur in the distillation column.

- Constituents of typical crude oil is shown below.

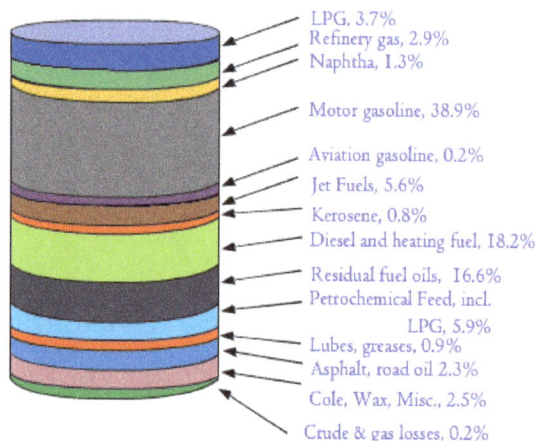

LPG, 3.7%
Refinery gas, 2.9%
Naphtha, 1.3%

Motor gasoline, 38.9%

Aviation gasoline, 0.2%
Jet Fuels, 5.6%
Kerosene, 0.8%
Diesel and heating fuel, 18.2%

Residual fuel oils, 16.6%
Petrochemical Feed, incl. LPG, 5.9%
Lubes, greases, 0.9%
Asphalt, road oil 2.3%
Cole, Wax, Misc., 2.5%
Crude & gas losses, 0.2%

Bomb Calorimeter

- Used to determine the calorific value of the liquid fuel.

- Liquid is burnt in the bomb in the presence of oxygen at about 2.5 MPa.

- The change in temperature in the water bath provides the calorific value of the fuel.

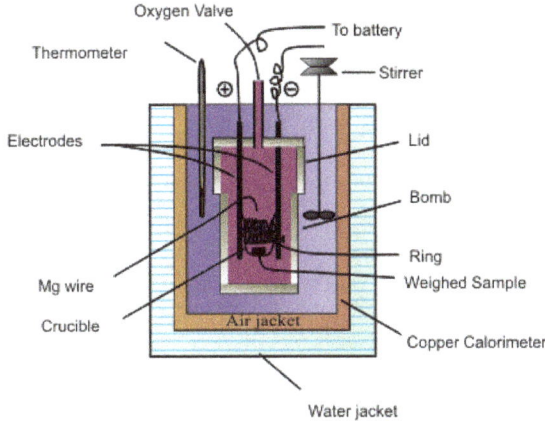

Oxygen Valve
Thermometer
To battery
Stirrer
Electrodes
Lid
Bomb
Ring
Weighed Sample
Mg wire
Crucible
Air jacket
Copper Calorimeter
Water jacket

Properties of Liquid Fuels

Specific Gravity: Ratio of mass density of fuel to mass density of water at the same temperature

$$SG = \frac{\rho_{fuel}}{\rho_{water}} \, (at \ the \ same \ temperature)$$

Reference temperature for fuel and water: 288.8 K

American Petroleum Institute (API) Scale:

$$APISG = \frac{141.5}{SG} - 131.4$$

Relation between APISG and HHV:

For Gasoline: $HHV = LHV + 93 \, (APISG - 10) kJ / kg$

For Kerosene: $HHV = LHV + 93 \, (APISG - 10) kJ / kg$

Auto Ignition Temperature : The lowest temperature required to make the combustion self sustained without any external aid

Flash Point: Minimum temperature at which liquid fuel will produce sufficient vapors to form a flammable mixture with air. Indicates maximum temperature at which liquid fuel can be stored without any fire hazard.

Fire Point: Minimum temperature at which liquid fuel produces sufficient vapors to form a flammable mixture with air that continuously supports combustion establishing flame instead of just flashing.

Smoke Point: Measure of the tendency of a liquid fuel to produce soot.

Properties of Common Liquid Fuels

Specific gravity	0.72 - 0.78	0.85	0.796	0.82	0.71
Kinematics viscosity @ 293 K (m²/s)	0.8×10^{-6}	2.5×10^{-6}	0.75×10^{-6}	3.626×10^{-6}	--
Boiling point range (K) @ STP	303 - 576	483 - 508	338	423-473	442
Flash point (K)	230	325	284	311	325
Auto ignition temperature (K)	643	527	737	483	--
Stoichiometric air/fuel by weight	14.7	14.7	6.45	15	15.1
Heat of Vaporization (kJ/kg)	380	375	1185	298.5	--
Lower heating value (MJ/kg)	43.5	45	20.1	45.2	43.3

Types of Fuels and Oxidizers

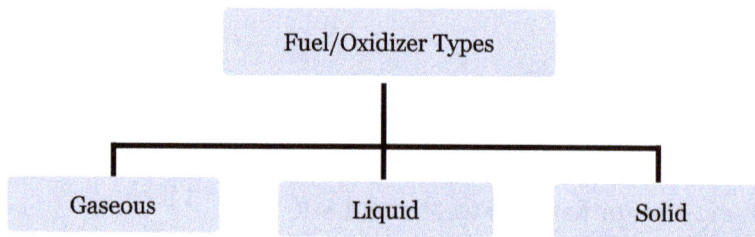

Gaseous Fuel and Oxidizer

Why gaseous fuels are preferred over liquid and solid fuels?

- Easier to control emissions.
- Easier to burn for higher efficiency.
- Gas handling system is less expensive.
- Commonly used gaseous fuels: CNG, LPG, Biogas, Producer gas, Coke oven gas, Acetylene, Methane, Hydrogen and Propane.

Types of Gaseous Fuel and Oxidizer

1	LPG	Air/O_2	Domestic Burner, Furnace
2	Natural Gas (NG)	Air/O_2	IC Engines, Furnaces
3	Producer Gas	Air/O_2	EC/IC Engines
4	CH_4, C_3H_8, H_2	Air/O_2	EC/IC Engines
5	Biogas	Air/O_2	EC/IC Engines, Burners
6	Acetylene	Air/O_2	Gas welding, Gas cutting
			* EC=External Combustion
			IC =Internal Combustion

Composition of Some Gaseous Fuels

Fuel	CO_2	O_2	N_2	CO	H_2	CH_4	C_2H_6	C_3H_8	C_4H_{10}
LPG	-	-	-	-	-	-	-	70	30
Natural Gas	-	-	5	-	-	90	5	-	-
Producer Gas	8	0.1	50	23.2	17.7	1	-	-	-
Propane	-	-	-	-	-	-	2.2	97.3	0.5
Biogas	33	-	1	-	1	65	-	-	-

Characterization of a Gaseous Fuel

Heating Value

- Amount of heat released per unit volume when it undergoes oxidation at normal pressure and temperature (0.1 MPa and 298 K).

- Lower heating value (LHV) – amount of heat released by burning 1 kg of fuel assuming the latent heat of vaporization in the reaction products is not recovered.

$$LHV = HHV - \frac{m_{H_2O}}{m_{fuel}} \Delta H_v$$

- Higher heating value (HHV) – heating value of the fuel when water is condensed.

- ΔH_v is the Latent heat of vaporization of water at 298.15 K

Junker's Calorimeter

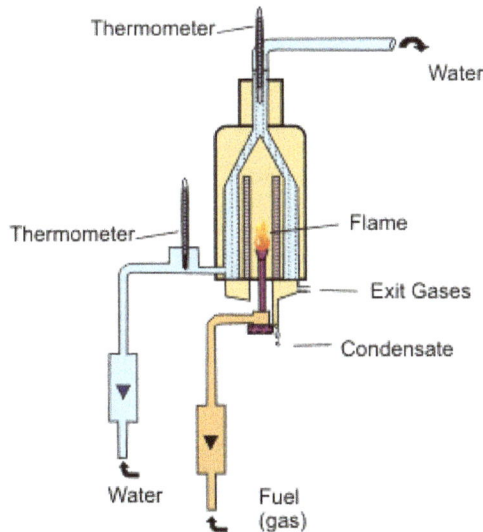

- Determines the heating value of the gaseous fuel.

- Fuel and air are burnt in a burner.

- Cooling water in the water jacket-absorbs the heat released during combustion.

- Heating value- calculated from the water flow rate and rise in temperature.

References

- Bradley, D (2009-06-25). "Combustion and the design of future engine fuels". Proceedings of the Institution of Mechanical Engineers, Part C: Journal of Mechanical Engineering Science. 223 (12): 2751–2765. doi:10.1243/09544062jmes1519

- Valorani, M.; Paolucci, S. (2009). "The G-Scheme: a framework for multi-scale adaptive model reduction". J. Comput. Phys. 228: 4665–4701. Bibcode:2009JCoPh.228.4665V. doi:10.1016/j.jcp.2009.03.011

- Chiavazzo, Eliodoro (2012). "Approximation of slow and fast dynamics in multiscale dynamical systems by the linearized Relaxation Redistribution Method". Journal of Computational Physics. 231: 1751–1765

- Shuttle-Mir History/Science/Microgravity/Candle Flame in Microgravity (CFM) – MGBX. Spaceflight.nasa.gov (1999-07-16). Retrieved on 2010-09-28

- Kooshkbaghi, Mahdi; Frouzakis, E. Christos; Chiavazzo, Eliodoro; Boulouchos, Konstantinos; Karlin, Ilya (2014). "The global relaxation redistribution method for reduction of combustion kinetics". The Journal of Chemical Physics. 141: 044102. Bibcode:2014JChPh.141d4102K. doi:10.1063/1.4890368

- Chiavazzo, Eliodoro; Karlin, Ilya; Gorban, Alexander (2010). "The role of thermodynamics in model reduction when using invariant grids". Commun. Comput. Phys. 8: 701–734. doi:10.4208/cicp.030709.210110a

- Dowling, A. P. (2000a). "Vortices, sound and flame – a damaging combination". The Aeronautical Journal of the RaeS

- Maas, U.; Pope, S.B. (1992). "Simplifying chemical kinetics: intrinsic low-dimensional manifolds in composition space". Combust. Flame. 88: 239–264. doi:10.1016/0010-2180(92)90034-m

2

An Overview of Thermodynamics Combustion

Thermodynamics is a sub-field of physics that deals with heat and temperature. The related elements of thermodynamics are guided by the four laws of thermodynamics. It can be classified into two parts: intensive and extensive. While the former can be understood in the context of pressure, temperature and spectral entropy, the latter's properties include mass, volume and enthalpy. Combustion is best understood in confluence with the major topics listed in the following chapter.

Thermodynamics of Combustion

Thermodynamic Properties

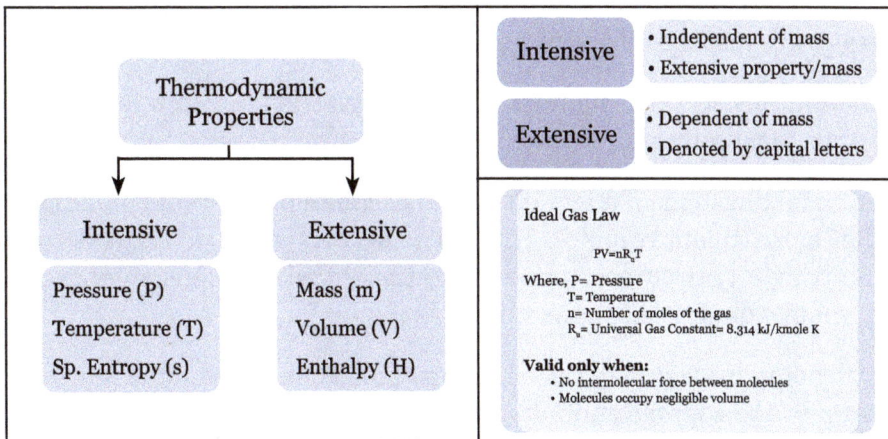

Thermodynamic Properties	
Intensive	• Independent of mass • Extensive property/mass
Extensive	• Dependent of mass • Denoted by capital letters

Intensive
- Pressure (P)
- Temperature (T)
- Sp. Entropy (s)

Extensive
- Mass (m)
- Volume (V)
- Enthalpy (H)

Ideal Gas Law

$$PV = nR_u T$$

Where, P= Pressure
T= Temperature
n= Number of moles of the gas
R_u= Universal Gas Constant= 8.314 kJ/kmole K

Valid only when:
• No intermolecular force between molecules
• Molecules occupy negligible volume

Gas Mixture

Dalton's Law	Total pressure of a gaseous mixture is the sum of the pressure which each component would exert if it alone occupies the same volume and temperature of mixture.
Gibb's Theorem	Internal energy of a mixture of ideal gases is equal to the mole/mass fraction weighed sums of the internal energy of individual components of mixture.

Gibbs-Dalton law: Employed to extract properties of mixture from individual gases.
Gibb's Theorem: Specific molar internal energy of the mixture from the constituent species.

Entropy of Mixing

In thermodynamics the entropy of mixing is the increase in the total entropy when several initially separate systems of different composition, each in a thermodynamic state of internal equilibrium, are mixed without chemical reaction by the thermodynamic operation of removal of impermeable partition(s) between them, followed by a time for establishment of a new thermodynamic state of internal equilibrium in the new unpartitioned closed system.

In general, the mixing may be constrained to occur under various prescribed conditions. In the customarily prescribed conditions, the materials are each initially at a common temperature and pressure, and the new system may change its volume, while being maintained at that same constant temperature, pressure, and chemical component masses. The volume available for each material to explore is increased, from that of its initially separate compartment, to the total common final volume. The final volume need not be the sum of the initially separate volumes, so that work can be done on or by the new closed system during the process of mixing, as well as heat being transferred to or from the surroundings, because of the maintenance of constant pressure and temperature.

The internal energy of the new closed system is equal to the sum of the internal energies of the initially separate systems. The reference values for the internal energies should be specified in a way that is constrained to make this so, maintaining also that the internal energies are respectively proportional to the masses of the systems.

For concision in this section, the term 'ideal material' is used to refer to an ideal gas (mixture) or an ideal solution.

In a process of mixing of ideal materials, the final common volume is the sum of the initial separate compartment volumes. There is no heat transfer and no work is done. The entropy of mixing is entirely accounted for by the diffusive expansion of each material into a final volume not initially accessible to it.

On the mixing of non-ideal materials, the total final common volume may be different from the sum of the separate initial volumes, and there may occur transfer of work or heat, to or from the surroundings; also there may be a departure of the entropy of mixing from that of the corresponding ideal case. That departure is the main reason for interest in entropy of mixing. These energy and entropy variables and their temperature dependences provide valuable information about the properties of the materials.

On a molecular level, the entropy of mixing is of interest because it is a macroscopic variable that provides information about constitutive molecular properties. In ideal materials, intermolecular forces are the same between every pair of molecular kinds, so that a molecule feels no difference between other molecules of its own kind and of those of the other kind. In non-ideal materials, there may be differences of intermolecular forces or specific molecular effects between different species, even though they are chemically non-reacting. The entropy of mixing provides information about constitutive differences of intermolecular forces or specific molecular effects in the materials.

The statistical concept of randomness is used for statistical mechanical explanation of the entropy of mixing. Mixing of ideal materials is regarded as random at a molecular level, and, correspondingly, mixing of non-ideal materials may be non-random.

Mixing of Ideal Materials at Constant Temperature and Pressure

In ideal materials, intermolecular forces are the same between every pair of molecular kinds, so that a molecule "feels" no difference between itself and its molecular neighbours. This is the reference case for examining corresponding mixings of non-ideal materials.

For example, two ideal gases, at the same temperature and pressure, are initially separated by a dividing partition.

Upon removal of the dividing partition, they expand into a final common volume (the sum of the two initial volumes), and the entropy of mixing $\Delta_{mix}S$ is given by

$$\Delta_{mix}S = -nR(x_1 \ln x_1 + x_2 \ln x_2).$$

where R is the gas constant, n the total number of moles and x_i the mole fraction of component i, which initially occupies volume $V_i = x_i V$. After the removal of the partition, the $n_i = nx_i$ moles of component i may explore the combined volume V, which causes an entropy increase equal to $nx_i R \ln(V / V_i) = -nRx_i \ln x_i$ for each component gas.

In this case, the increase in entropy is due entirely to the irreversible processes of expansion of the two gases, and involves no heat or work flow between the system and its surroundings.

Gibbs Free Energy of Mixing

The Gibbs free energy change $\Delta_{mix}G = \Delta_{mix}H - T\Delta_{mix}S$ determines whether mixing at constant (absolute) temperature T and pressure p is a spontaneous process. This quantity combines two physical effects—the enthalpy of mixing, which is a measure of the energy change, and the entropy of mixing considered here.

For an ideal gas mixture or an ideal solution, there is no enthalpy of mixing ($\Delta_{mix}H$), so that the Gibbs free energy of mixing is given by the entropy term only:

$$\Delta_{mix}G = -T\Delta_{mix}S$$

For an ideal solution, the Gibbs free energy of mixing is always negative, meaning that mixing of ideal solutions is always spontaneous. The lowest value is when the mole fraction is 0.5 for a mixture of two components, or 1/n for a mixture of n components.

Solutions and Temperature Dependence of Miscibility

Ideal and Regular Solutions

The above equation for the entropy of mixing of ideal gases is valid also for certain liquid (or solid) solutions—those formed by completely random mixing so that the components move independently in the total volume. Such random mixing of solutions occurs if the interaction energies between unlike molecules are similar to the average interaction energies between like molecules. The value of the entropy corresponds exactly to random mixing for ideal solutions and for regular solutions, and approximately so for many real solutions.

For binary mixtures the entropy of random mixing can be considered as a function of the mole fraction of one component.

$$\Delta_{mix}S = -nR(x_1 \ln x_1 + x_2 \ln x_2) = -nR[x \ln x + (1-x) \ln(1-x)]$$

For all possible mixtures, $0 < x < 1$, so that $\ln x$ and $\ln(1-x)$ are both negative and the entropy of mixing $\Delta_{mix}S$ is positive and favors mixing of the pure components.

Also the curvature of $\Delta_{mix}S$ as a function of x is given by the second derivative

$$\left(\frac{\partial^2 \Delta_{mix}S}{\partial x^2} \right)_{T,P} = -nR\left(\frac{1}{x} + \frac{1}{1-x} \right)$$

This curvature is negative for all possible mixtures $(0 < x < 1)$, so that mixing two solutions to form a solution of intermediate composition also increases the entropy of the system. Random mixing therefore always favors miscibility and opposes phase separation.

For ideal solutions, the enthalpy of mixing is zero so that the components are miscible in all proportions. For regular solutions a positive enthalpy of mixing may cause incomplete miscibility (phase separation for some compositions) at temperatures below the upper critical solution temperature (UCST). This is the minimum temperature at which the $-T\Delta_{mix}S$ term in the Gibbs energy of mixing is sufficient to produce miscibility in all proportions.

Systems with a Lower Critical Solution Temperature

Nonrandom mixing with a lower entropy of mixing can occur when the attractive interactions between unlike molecules are significantly stronger (or weaker) than the mean interactions between like molecules. For some systems this can lead to a lower critical solution temperature (LCST) or lower limiting temperature for phase separation.

For example, triethylamine and water are miscible in all proportions below 19°C, but above this critical temperature, solutions of certain compositions separate into two phases at equilibrium with each other. This means that $\Delta_{mix}G$ is negative for mixing of the two phases below 19°C and positive above this temperature. Therefore,

$\Delta_{mix}S = -\left(\dfrac{\partial \Delta_{mix}G}{\partial T}\right)_P$ is negative for mixing of these two equilibrium phases. This is due to

the formation of attractive hydrogen bonds between the two components that prevent random mixing. Triethylamine molecules cannot form hydrogen bonds with each other but only with water molecules, so in solution they remain associated to water molecules with loss of entropy. The mixing that occurs below 19°C is due not to entropy but to the enthalpy of formation of the hydrogen bonds.

Lower critical solution temperatures also occur in many polymer-solvent mixtures. For polar systems such as polyacrylic acid in 1,4-dioxane, this is often due to the formation of hydrogen bonds between polymer and solvent. For nonpolar systems such as polystyrene in cyclohexane, phase separation has been observed in sealed tubes (at high pressure) at temperatures approaching the liquid-vapor critical point of the solvent. At such temperatures the solvent expands much more rapidly than the polymer, whose segments are covalently linked. Mixing therefore requires contraction of the solvent for compatibility of the polymer, resulting in a loss of entropy.

Statistical Thermodynamical Explanation of the Entropy of Mixing of Ideal Gases

Since thermodynamic entropy can be related to statistical mechanics or to information theory, it is possible to calculate the entropy of mixing using these two approaches. Here we consider the simple case of mixing ideal gases.

Proof from Statistical Mechanics

Assume that the molecules of two different substances are approximately the same size, and regard space as subdivided into a square lattice whose cells are the size of the molecules. (In fact, any lattice would do, including close packing.) This is a crystal-like conceptual model to identify the molecular centers of mass. If the two phases are liquids, there is no spatial uncertainty in each one individually. (This is, of course, an approximation. Liquids have a "free volume". This is why they are (usually) less dense than solids.) Everywhere we look in component 1, there is a molecule present, and likewise for component 2. After the two different substances are intermingled (assuming they are miscible), the liquid is still dense with molecules, but now there is uncertainty about what kind of molecule is in which location. Of course, any idea of identifying molecules in given locations is a thought experiment, not something one could do, but the calculation of the uncertainty is well-defined.

We can use Boltzmann's equation for the entropy change as applied to the mixing process

$$\Delta_{mix}S = k_B \ln \Omega$$

where k_B is Boltzmann's constant. We then calculate the number of ways Ω of arranging N_1 molecules of component 1 and N_2 molecules of component 2 on a lattice, where

$$N = N_1 + N_2$$

is the total number of molecules, and therefore the number of lattice sites. Calculating the number of permutations of   objects, correcting for the fact that N_1 of them are *identical* to one another, and likewise for N_2,

$$\Omega = N! / N_1! N_2!$$

After applying Stirling's approximation for the factorial of a large integer m:

$$\ln m! = \sum_k \ln k \approx \int_1^m dk \ln k = m \ln m - m,$$

the result is

$$\Delta_{mix}S = -k_B[N_1 \ln(N_1 / N) + N_2 \ln(N_2 / N)] = -k_B N[x_1 \ln x_1 + x_2 \ln x_2]$$

where we have introduced the mole fractions, which are also the probabilities of finding any particular component in a given lattice site.

$$x_1 = N_1 / N = p_1 \text{ and } x_2 = N_2 / N = p_2$$

Since the Boltzmann constant $k_B = R / N_A$, where N_A is Avogadro's number, and the number of molecules $N = nN_A$, we recover the thermodynamic expression for the mixing of two ideal gases,

$$\Delta_{mix}S = -nR[x_1 \ln x_1 + x_2 \ln x_2]$$

This expression can be generalized to a mixture of r components, N_i , with $i = 1, 2, 3, \ldots, r$

$$\Delta_{mix}S = -k_B \sum_{i=1}^r N_i \ln(N_i / N) = -Nk_B \sum_{i=1}^r x_i \ln x_i = -nR \sum_{i=1}^r x_i \ln x_i$$

Relationship to Information Theory

The entropy of mixing is also proportional to the Shannon entropy or compositional uncertainty of information theory, which is defined without requiring Stirling's approximation. Claude Shannon introduced this expression for use in information theory, but similar formulas can be found as far back as the work of Ludwig Boltzmann and J. Willard Gibbs. The Shannon uncertainty is not the same as the Heisenberg uncertainty principle in quantum mechanics which is based on variance. The Shannon entropy is defined as:

$$H = -\sum_{i=1}^{r} p_i \ln(p_i)$$

where p_i is the probability that an information source will produce the i th symbol from an r-symbol alphabet and is independent of previous symbols. (thus i runs from 1 to r). H is then a measure of the expected amount of information ($\log p_i$) missing before the symbol is known or measured, or, alternatively, the expected amount of information supplied when the symbol becomes known. The set of messages of length N symbols from the source will then have an entropy of N*H.

The thermodynamic entropy is only due to positional uncertainty, so we may take the "alphabet" to be any of the r different species in the gas, and, at equilibrium, the probability that a given particle is of type i is simply the mole fraction x_i for that particle. Since we are dealing with ideal gases, the identity of nearby particles is irrelevant. Multiplying by the number of particles N yields the change in entropy of the entire system from the unmixed case in which all of the p_i were either 1 or 0. We again obtain the entropy of mixing on multiplying by the Boltzmann constant k_B .

$$\Delta_{mix}S = -Nk_B \sum_{i=1}^{r} x_i \ln x_i$$

So thermodynamic entropy with "r" chemical species with a total of N particles has a parallel to an information source that has "r" distinct symbols with messages that are N symbols long.

Application to Gases

In gases there is a lot more spatial uncertainty because most of their volume is merely empty space. We can regard the mixing process as allowing the contents of the two originally separate contents to expand into the combined volume of the two conjoined containers. The two lattices that allow us to conceptually localize molecular centers of mass also join. The total number of empty cells is the sum of the numbers of empty cells in the two components prior to mixing. Consequently, that part of the spatial uncertainty concerning whether *any* molecule is present in a lattice cell is the sum of the initial values, and does not increase upon "mixing".

Almost everywhere we look, we find empty lattice cells. Nevertheless, we do find molecules in a few occupied cells. When there is real mixing, for each of those few occupied cells, there is a contingent uncertainty about which kind of molecule it is. When there is no real mixing because the two substances are identical, there is no uncertainty about which kind of molecule it is. Using conditional probabilities, it turns out that the analytical problem for the small subset of occupied cells is exactly the same as for mixed liquids, and the *increase* in the entropy, or spatial uncertainty, has exactly the same form as obtained previously. Obviously the subset of occupied cells is not the same at different times. But only when there is real mixing and an occupied cell is found do we ask which kind of molecule is there.

Application to Solutions

If the solute is a crystalline solid, the argument is much the same. A crystal has no spatial uncertainty at all, except for crystallographic defects, and a (perfect) crystal allows us to localize the molecules using the crystal symmetry group. The fact that volumes do not add when dissolving a solid in a liquid is not important for condensed phases. If the solute is not crystalline, we can still use a spatial lattice, as good an approximation for an amorphous solid as it is for a liquid.

The Flory–Huggins solution theory provides the entropy of mixing for polymer solutions, in which the macromolecules are huge compared to the solvent molecules. In this case, the assumption is made that each monomer subunit in the polymer chain occupies a lattice site.

Note that solids in contact with each other also slowly interdiffuse, and solid mixtures of two or more components may be made at will (alloys, semiconductors, etc.). Again, the same equations for the entropy of mixing apply, but only for homogeneous, uniform phases.

Mixing Under Other Constraints

Mixing with and without Change of Available Volume

In the established customary usage, expressed in the lead section of this article, the entropy of mixing comes from two mechanisms, the intermingling and possible interactions of the distinct molecular species, and the change in the volume available for each molecular species, or the change in concentration of each molecular species. For ideal gases, the entropy of mixing at prescribed common temperature and pressure has nothing to do with mixing in the sense of intermingling and interactions of molecular species, but is only to do with expansion into the common volume.

According to Fowler and Guggenheim (1939/1965), the conflating of the just-mentioned two mechanisms for the entropy of mixing is well established in customary terminology, but can be confusing unless it is borne in mind that the independent variables are the common initial and final temperature and total pressure; if the respective partial pressures or the total volume are chosen as independent variables instead of the total pressure, the description is different.

1. Mixing with Each Gas Kept at Constant Partial Volume, with Changing Total Volume:

In contrast to the established customary usage, "mixing" might be conducted reversibly at constant volume for each of two fixed masses of gases of equal volume, being mixed by gradually merging their initially separate volumes by use of two ideal semipermeable membranes, each permeable only to one of the respective gases, so that the respective volumes available to each gas remain constant during the merge. Either one of the common temperature or the common pressure is chosen to be independently controlled by the experimenter, the other being allowed to vary so as to maintain constant volume for each mass of gas. In this kind of "mixing", the final common volume is equal to each of the respective separate initial volumes, and each gas finally occupies the same volume as it did initially.

This constant volume kind of "mixing", in the special case of perfect gases, is referred to in what is sometimes called Gibbs' theorem. It states that the entropy of such "mixing" of perfect gases is zero.

2. Mixing at Constant Total Volume and Changing Partial Volumes, with Mechanically Controlled Varying Pressure, and Constant Temperature:

An experimental demonstration may be considered. The two distinct gases, in a cylinder of constant total volume, are at first separated by two contiguous pistons made respectively of two suitably specific ideal semipermeable membranes. Ideally slowly and fictively reversibly, at constant temperature, the gases are allowed to mix in the volume between the separating membranes, forcing them apart, thereby supplying work to an external system. The energy for the work comes from the heat reservoir that keeps the temperature constant. Then, by externally forcing ideally slowly the separating membranes together, back to contiguity, work is done on the mixed gases, fictively reversibly separating them again, so that heat is returned to the heat reservoir at constant temperature. Because the mixing and separation are ideally slow and fictively reversible, the work supplied by the gases as they mix is equal to the work done in separating them again. Passing from fictive reversibility to physical reality, some amount of additional work, that remains external to the gases and the heat reservoir, must be provided from an external source for this cycle, as required by the second law of thermodynamics, because this cycle has only one heat reservoir at constant temperature, and the external provision of work cannot be completely efficient.

Gibbs' Paradox

- "Mixing" of Identical Species Versus Mixing of Closely Similar But non-identical Species:

For entropy of mixing to exist, the putatively mixing molecular species must be chemically or physically detectably distinct. Thus arises the so-called *Gibbs paradox*, as follows. If molecular species are identical, there is no entropy change on mixing them, because, defined in thermodynamic terms, there is no mass transfer, and thus no thermodynamically recognized process of mixing. Yet the slightest detectable difference in constitutive properties between the two species yields a thermodynamically recognized process of transfer with mixing, and a possibly considerable entropy change, namely the entropy of mixing.

The "paradox" arises because any detectable constitutive distinction, no matter how slight, can lead to a considerably large change in amount of entropy as a result of mixing. Though a continuous change in the properties of the materials that are mixed might make the degree of constitutive difference tend continuously to zero, the entropy change would nonetheless vanish discontinuously when the difference reached zero.

From a general physical viewpoint, this discontinuity is paradoxical. But from a specifically thermodynamic viewpoint, it is not paradoxical, because in that discipline the degree of constitutive difference is not questioned; it is either there or not there. Gibbs himself did not see it as paradoxical. Distinguishability of two materials is a constitutive, not a thermodynamic, difference, for the laws of thermodynamics are the same for every material, while their constitutive characteristics are diverse.

Though one might imagine a continuous decrease of the constitutive difference between any two chemical substances, physically it cannot be continuously decreased till it actually vanishes. It is hard to think of a smaller difference than that between ortho- and para-hydrogen. Yet they differ by a finite amount. The hypothesis, that the distinction might tend continuously to zero, is unphysical. This is neither examined nor explained by thermodynamics. Differences of constitution are explained by quantum mechanics, which postulates discontinuity of physical processes.

For a detectable distinction, some means should be physically available. One theoretical means would be through an ideal semi-permeable membrane. It should allow passage, backwards and forwards, of one species, while passage of the other is prevented entirely. The entirety of prevention should include perfect efficacy over a practically infinite time, in view of the nature of thermodynamic equilibrium. Even the slightest departure from ideality, as assessed over a finite time, would extend to utter non-ideality, as assessed over a practically infinite time. Such quantum phenomena as tunneling ensure that nature does not allow such membrane ideality as would support the theoretically demanded continuous decrease, to zero, of detectable distinction. The decrease to zero detectable distinction must be discontinuous.

For ideal gases, the entropy of mixing does not depend on the degree of difference between the distinct molecular species, but only on the fact that they are distinct; for non-ideal gases, the entropy of mixing can depend on the degree of difference of the distinct molecular species. The suggested or putative "mixing" of identical molecular species is not in thermodynamic terms a mixing at all, because thermodynamics refers to states specified by state variables, and does not permit an imaginary labelling of particles. Only if the molecular species are different is there mixing in the thermodynamic sense.

Thermodynamics

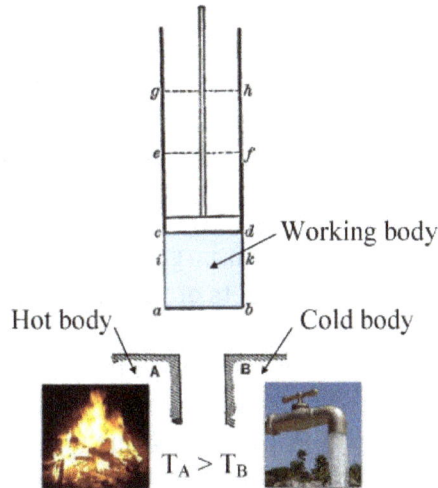

Annotated color version of the original 1824 Carnot heat engine showing the hot body (boiler), working body (system, steam), and cold body (water), the letters labeled according to the stopping points in Carnot cycle.

Thermodynamics is a branch of physics concerned with heat and temperature and their relation to energy and work. The behavior of these quantities is governed by the four laws of thermodynamics, irrespective of the composition or specific properties of the material or system in question. The laws of thermodynamics are explained in terms of microscopic constituents by statistical mechanics. Thermodynamics applies to a wide variety of topics in science and engineering, especially physical chemistry, chemical engineering and mechanical engineering.

Historically, thermodynamics developed out of a desire to increase the efficiency of early steam engines, particularly through the work of French physicist Nicolas Léonard Sadi Carnot (1824) who believed that engine efficiency was the key that could help France win the Napoleonic Wars. Scottish physicist Lord Kelvin was the first to formulate a concise definition of thermodynamics in 1854 which stated, *"Thermo-dynamics is the subject of the relation of heat to forces acting between contiguous parts of bodies, and the relation of heat to electrical agency."*

The initial application of thermodynamics to mechanical heat engines was extended early on to the study of chemical compounds and chemical reactions. Chemical thermodynamics studies the nature of the role of entropy in the process of chemical reactions and has provided the bulk of expansion and knowledge of the field. Other formulations of thermodynamics emerged in the following decades. Statistical thermodynamics, or statistical mechanics, concerned itself with statistical predictions of the collective motion of particles from their microscopic behavior. In 1909, Constantin Carathéodory presented a purely mathematical approach to the field in his axiomatic formulation of thermodynamics, a description often referred to as *geometrical thermodynamics*.

A description of any thermodynamic system employs the four laws of thermodynamics that form an axiomatic basis. The first law specifies that energy can be exchanged between physical systems as heat and work. The second law defines the existence of a quantity called entropy, that describes the direction, thermodynamically, that a system can evolve and quantifies the state of order of a system and that can be used to quantify the useful work that can be extracted from the system.

In thermodynamics, interactions between large ensembles of objects are studied and categorized. Central to this are the concepts of the thermodynamic *system* and its *surroundings*. A system is composed of particles, whose average motions define its properties, and those properties are in turn are related to one another through equations of state. Properties can be combined to express internal energy and thermodynamic potentials, which are useful for determining conditions for equilibrium and spontaneous processes.

École Polytechnique	Glasgow school	Berlin school	Edinburgh school
Sadi Carnot (1796-1832)	William Thomson (1824-1907)	Rudolf Clausius (1822-1888)	James Maxwell (1831-1879)
Vienna school	Gibbsian school	Dresden school	Dutch school
Ludwig Boltzmann (1844-1906)	Willard Gibbs (1839-1903)	Gustav Zeuner (1828-1907)	Johannes der Waals (1837-1923)

The thermodynamicists representative of the original eight founding schools of thermodynamics. The schools with the most-lasting effect in founding the modern versions of thermodynamics are the Berlin school, particularly as established in Rudolf Clausius's 1865 textbook. *The Mechanical Theory of Heat*, the Vienna school, with the statistical mechanics of Ludwig Boltzmann, and the Gibbsian school at Yale University, American engineer Willard Gibbs' 1876 *On the Equilibrium of Heterogeneous Substances* launching chemical thermodynamics.

With these tools, thermodynamics can be used to describe how systems respond to changes in their environment. This can be applied to a wide variety of topics in science and engineering, such as engines, phase transitions, chemical reactions, transport phenomena, and even black holes. The results of thermodynamics are essential for other fields of physics and for chemistry, chemical engineering, aerospace engineering, mechanical engineering, cell biology, biomedical engineering, materials science, and economics, to name a few.

This section is focused mainly on classical thermodynamics which primarily studies systems in thermodynamic equilibrium. Non-equilibrium thermodynamics is often treated as an extension of the classical treatment, but statistical mechanics has brought many advances to that field.

History

The history of thermodynamics as a scientific discipline generally begins with Otto von Guericke who, in 1650, built and designed the world's first vacuum pump and demonstrated a vacuum using his Magdeburg hemispheres. Guericke was driven to make a vacuum in order to disprove Aristotle's long-held supposition that 'nature abhors a vacuum'. Shortly after Guericke, the English physicist and chemist Robert Boyle had learned of Guericke's designs and, in 1656, in coordination with English scientist Robert Hooke, built an air pump. Using this pump, Boyle and Hooke noticed a correlation between pressure, temperature, and volume. In time, Boyle's Law was formulated, which states that pressure and volume are inversely proportional. Then, in 1679, based on these concepts, an associate of Boyle's named Denis Papin built a steam digester, which was a closed vessel with a tightly fitting lid that confined steam until a high pressure was generated.

Later designs implemented a steam release valve that kept the machine from exploding. By watching the valve rhythmically move up and down, Papin conceived of the idea of a piston and a cylinder engine. He did not, however, follow through with his design. Nevertheless, in 1697, based on Papin's designs, engineer Thomas Savery built the first engine, followed by Thomas Newcomen in 1712. Although these early engines were crude and inefficient, they attracted the attention of the leading scientists of the time.

The fundamental concepts of heat capacity and latent heat, which were necessary for the development of thermodynamics, were developed by Professor Joseph Black at the University of Glasgow, where James Watt was employed as an instrument maker. Black and Watt performed experiments together, but it was Watt who conceived the idea of the external condenser which resulted in a large increase in steam engine efficiency. Drawing on all the previous work led Sadi Carnot, the "father of thermodynamics", to publish *Reflections on the Motive Power of Fire* (1824), a discourse on heat, power, energy and engine efficiency. The paper outlined the basic energetic relations between the Carnot engine, the Carnot cycle, and motive power. It marked the start of thermo-dynamics as a modern science.

The first thermodynamic textbook was written in 1859 by William Rankine, originally trained as a physicist and a civil and mechanical engineering professor at the University of Glasgow. The first and second laws of thermodynamics emerged simultaneously in the 1850s, primarily out of the works of William Rankine, Rudolf Clausius, and William Thomson (Lord Kelvin).

The foundations of statistical thermodynamics were set out by physicists such as James Clerk Maxwell, Ludwig Boltzmann, Max Planck, Rudolf Clausius and J. Willard Gibbs.

During the years 1873-76 the American mathematical physicist Josiah Willard Gibbs published a series of three papers, the most famous being *On the Equilibrium of Heterogeneous Substances*, in which he showed how thermodynamic processes, including chemical reactions, could be graphically analyzed, by studying the energy, entropy, volume, temperature and pressure of the thermodynamic system in such a manner, one can determine if a process would occur spontaneously. Also Pierre Duhem in the 19th century wrote about chemical thermodynamics. During the early 20th century, chemists such as Gilbert N. Lewis, Merle Randall, and E. A. Guggenheim applied the mathematical methods of Gibbs to the analysis of chemical processes.

Etymology

The etymology of *thermodynamics* has an intricate history. It was first spelled in a hy-phenated form as an adjective (*thermo-dynamic*) and from 1854 to 1868 as the noun *thermo-dynamics* to represent the science of generalized heat engines.

American biophysicist Donald Haynie claims that *thermodynamics* was coined in 1840 from the Greek root *therme,* meaning heat and *dynamis,* meaning power. However, this etymology has been cited as unlikely.

Pierre Perrot claims that the term *thermodynamics* was coined by James Joule in 1858 to designate the science of relations between heat and power, however, Joule never used that term, but used instead the term *perfect thermo-dynamic engine* in reference to Thomson's 1849 phraseology.

By 1858, *thermo-dynamics*, as a functional term, was used in William Thomson's paper *An Account of Carnot's Theory of the Motive Power of Heat*.

Branches of Description

The study of thermodynamical systems has developed into several related branches, each using a different fundamental model as a theoretical or experimental basis, or applying the principles to varying types of systems.

Classical Thermodynamics

Classical thermodynamics is the description of the states of thermodynamic systems at near-equilibrium, that uses macroscopic, measurable properties. It is used to model exchanges of energy, work and heat based on the laws of thermodynamics. The qualifier *classical* reflects the fact that it represents the first level of understanding of the subject as it developed in the 19th century and describes the changes of a system in terms of macroscopic empirical (large scale, and measurable) parameters. A microscopic interpretation of these concepts was later provided by the development of *statistical mechanics.*

Statistical Mechanics

Statistical mechanics, also called statistical thermodynamics, emerged with the development of atomic and molecular theories in the late 19th century and early 20th century, and supplemented classical thermodynamics with an interpretation of the microscopic interactions between individual particles or quantum-mechanical states. This field relates the microscopic properties of individual atoms and molecules to the macroscopic, bulk properties of materials that can be observed on the human scale, thereby explaining classical thermodynamics as a natural result of statistics, classical mechanics, and quantum theory at the microscopic level.

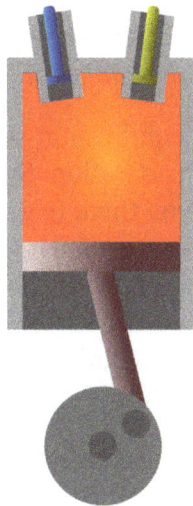

Chemical Thermodynamics

Chemical thermodynamics is the study of the interrelation of energy with chemical reactions or with a physical change of state within the confines of the laws of thermodynamics.

Treatment of Equilibrium

Equilibrium thermodynamics is the systematic study of transformations of matter and energy in systems as they approach equilibrium. The word equilibrium implies a state of balance. In an equilibrium state there are no unbalanced potentials, or driving forces, within the system. A central aim in equilibrium thermodynamics is: given a system in a well-defined initial state, subject to accurately specified constraints, to calculate what the state of the system will be once it has reached equilibrium.

Non-equilibrium thermodynamics is a branch of thermodynamics that deals with systems that are not in thermodynamic equilibrium. Most systems found in nature are not in thermodynamic equilibrium because they are not in stationary states, and are continuously and discontinuously subject to flux of matter and energy to and from other systems. The thermodynamic study of non-equilibrium systems requires more general concepts than are dealt with by equilibrium thermodynamics. Many natural systems still today remain beyond the scope of currently known macroscopic thermodynamic methods.

Laws of Thermodynamics

Thermodynamics is principally based on a set of four laws which are universally valid when applied to systems that fall within the constraints implied by each. In the various theoretical descriptions of thermodynamics these laws may be expressed in seemingly differing forms, but the most prominent formulations are the following:

- Zeroth law of thermodynamics: *If two systems are each in thermal equilibrium with a third, they are also in thermal equilibrium with each other.*

This statement implies that thermal equilibrium is an equivalence relation on the set of thermodynamic systems under consideration. Systems are said to be in equilibrium if the small, random exchanges between them (e.g. Brownian motion) do not lead to a net change in energy. This law is tacitly assumed in every measurement of temperature. Thus, if one seeks to decide if two bodies are at the same temperature, it is not necessary to bring them into contact and measure any changes of their observable properties in time. The law provides an empirical definition of temperature and justification for the construction of practical thermometers.

The zeroth law was not initially recognized as a law, as its basis in thermodynamical equilibrium was implied in the other laws. The first, second, and third laws had been explicitly stated prior and found common acceptance in the physics community. Once the importance of the zeroth law for the definition of temperature was realized, it was impracticable to renumber the other laws, hence it was numbered the *zeroth law*.

- First law of thermodynamics: *The internal energy of an isolated system is constant.*

The first law of thermodynamics is an expression of the principle of conservation of energy. It states that energy can be transformed (changed from one form to another), but cannot be created or destroyed.

The first law is usually formulated by saying that the change in the internal energy of a closed thermodynamic system is equal to the difference between the heat supplied to the system and the amount of work done by the system on its surroundings. It is important to note that internal energy is a state of the system whereas heat and work modify the state of the system. In other words, an change of internal energy of a system may be achieved by any combination of heat and work added or removed from the system as long as those total to the change of internal energy. The manner by which a system achieves its internal energy is path independent.

- Second law of thermodynamics: *Heat cannot spontaneously flow from a colder location to a hotter location.*

The second law of thermodynamics is an expression of the universal principle of decay observable in nature. The second law is an observation of the fact that over time, differences in temperature, pressure, and chemical potential tend to even out in a physical system that is isolated from the outside world. Entropy is a measure of how much this process has progressed. The entropy of an isolated system which is not in equilibrium will tend to increase over time, approaching a maximum value at equilibrium.

In classical thermodynamics, the second law is a basic postulate applicable to any system involving heat energy transfer; in statistical thermodynamics, the second law is a consequence of the assumed randomness of molecular chaos. There are many versions of the second law, but they all have the same effect, which is to explain the phenomenon of irreversibility in nature.

- Third law of thermodynamics: *As a system approaches absolute zero, all processes cease and the entropy of the system approaches a minimum value.*

The third law of thermodynamics is a statistical law of nature regarding entropy and the impossibility of reaching absolute zero of temperature. This law provides an absolute reference point for the determination of entropy. The entropy determined relative to this point is the absolute entropy. Alternate definitions are, "the entropy of all systems and of all states of a system is smallest at absolute zero," or equivalently "it is impossible to reach the absolute zero of temperature by any finite number of processes".

Absolute zero, at which all activity would stop if it were possible to happen, is $-273.15°C$ (degrees Celsius), or $-459.67°F$ (degrees Fahrenheit) or 0 K (kelvin).

System Models

An important concept in thermodynamics is the thermodynamic system, which is a precisely defined region of the universe under study. Everything in the universe except the system is called the *surroundings*. A system is separated from the remainder of the universe by a *boundary* which may be a physical boundary or notional, but which by convention defines a finite volume. Exchanges of work, heat, or matter between the system and the surroundings take place across this boundary.

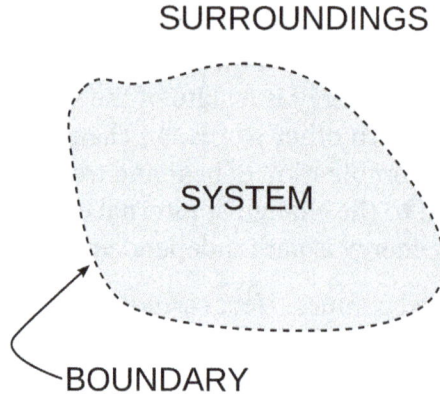

SURROUNDINGS

SYSTEM

BOUNDARY

A diagram of a generic thermodynamic system

In practice, the boundary of a system is simply an imaginary dotted line drawn around a volume within which is going to be a change in the internal energy of that volume. Anything that passes across the boundary that effects a change in the internal energy of the system needs to be accounted for in the energy balance equation. The volume can be the region surrounding a single atom resonating energy, such as Max Planck defined in 1900; it can be a body of steam or air in a steam engine, such as Sadi Carnot defined in 1824; it can be the body of a tropical cyclone, such as Kerry Emanuel theorized in 1986 in the field of atmospheric thermodynamics; it could also be just one nuclide (i.e. a system of quarks) as hypothesized in quantum thermodynamics, or the event horizon of a black hole.

Boundaries are of four types: fixed, movable, real, and imaginary. For example, in an engine, a fixed boundary means the piston is locked at its position, within which a constant volume process might occur. If the piston is allowed to move that boundary is movable while the cylinder and cylinder head boundaries are fixed. For closed systems, boundaries are real while for open systems boundaries are often imaginary. In the case of a jet engine, a fixed imaginary boundary might be assumed at the intake of the engine, fixed boundaries along the surface of the case and a second fixed imaginary boundary across the exhaust nozzle.

Generally, thermodynamics distinguishes three classes of systems, defined in terms of what is allowed to cross their boundaries:

Interactions of Thermodynamic Systems			
Type of system	Mass flow	Work	Heat
Open	✓	✓	✓
Closed	✗	✓	✓

Thermally isolated	✗	✓	✗
Mechanically isolated	✗	✗	✓
Isolated	✗	✗	✗

As time passes in an isolated system, internal differences of pressures, densities, and temperatures tend to even out. A system in which all equalizing processes have gone to completion is said to be in a state of thermodynamic equilibrium.

Once in thermodynamic equilibrium, a system's properties are, by definition, unchanging in time. Systems in equilibrium are much simpler and easier to understand than are systems which are not in equilibrium. Often, when analysing a dynamic thermodynamic process, the simplifying assumption is made that each intermediate state in the process is at equilibrium, producing thermodynamic processes which develop so slowly as to allow each intermediate step to be an equilibrium state and are said to be reversible processes.

States and Processes

When a system is at equilibrium under a given set of conditions, it is said to be in a definite thermodynamic state. The state of the system can be described by a number of state quantities that do not depend on the process by which the system arrived at its state. They are called intensive variables or extensive variables according to how they change when the size of the system changes. The properties of the system can be described by an equation of state which specifies the relationship between these variables. State may be thought of as the instantaneous quantitative description of a system with a set number of variables held constant.

A thermodynamic process may be defined as the energetic evolution of a thermodynamic system proceeding from an initial state to a final state. It can be described by process quantities. Typically, each thermodynamic process is distinguished from other processes in energetic character according to what parameters, such as temperature, pressure, or volume, etc., are held fixed. Furthermore, it is useful to group these processes into pairs, in which each variable held constant is one member of a conjugate pair.

Several commonly studied thermodynamic processes are:

- Adiabatic process: occurs without loss or gain of energy by heat
- Isenthalpic process: occurs at a constant enthalpy
- Isentropic process: a reversible adiabatic process, occurs at a constant entropy
- Isobaric process: occurs at constant pressure
- Isochoric process: occurs at constant volume (also called isometric/isovolumetric)

- Isothermal process: occurs at a constant temperature

- Steady state process: occurs without a change in the internal energy

Instrumentation

There are two types of thermodynamic instruments, the meter and the reservoir. A thermodynamic meter is any device which measures any parameter of a thermodynamic system. In some cases, the thermodynamic parameter is actually defined in terms of an idealized measuring instrument. For example, the zeroth law states that if two bodies are in thermal equilibrium with a third body, they are also in thermal equilibrium with each other. This principle, as noted by James Maxwell in 1872, asserts that it is possible to measure temperature. An idealized thermometer is a sample of an ideal gas at constant pressure. From the ideal gas law $pV=nRT$, the volume of such a sample can be used as an indicator of temperature; in this manner it defines temperature. Although pressure is defined mechanically, a pressure-measuring device, called a barometer may also be constructed from a sample of an ideal gas held at a constant temperature. A calorimeter is a device which is used to measure and define the internal energy of a system.

A thermodynamic reservoir is a system which is so large that its state parameters are not appreciably altered when it is brought into contact with the system of interest. When the reservoir is brought into contact with the system, the system is brought into equilibrium with the reservoir. For example, a pressure reservoir is a system at a particular pressure, which imposes that pressure upon the system to which it is mechanically connected. The Earth's atmosphere is often used as a pressure reservoir. If ocean water is used to cool a power plant, the ocean is often a temperature reservoir in the analysis of the power plant cycle.

Conjugate Variables

The central concept of thermodynamics is that of energy, the ability to do work. By the First Law, the total energy of a system and its surroundings is conserved. Energy may be transferred into a system by heating, compression, or addition of matter, and extracted from a system by cooling, expansion, or extraction of matter. In mechanics, for example, energy transfer equals the product of the force applied to a body and the resulting displacement.

Conjugate variables are pairs of thermodynamic concepts, with the first being akin to a "force" applied to some thermodynamic system, the second being akin to the resulting "displacement," and the product of the two equalling the amount of energy transferred. The common conjugate variables are:

- Pressure-volume (the mechanical parameters);

- Temperature-entropy (thermal parameters);

- Chemical potential-particle number (material parameters).

Potentials

Thermodynamic potentials are different quantitative measures of the stored energy in a system. Potentials are used to measure energy changes in systems as they evolve from an initial state to a final state. The potential used depends on the constraints of the system, such as constant temperature or pressure. For example, the Helmholtz and Gibbs energies are the energies available in a system to do useful work when the temperature and volume or the pressure and temperature are fixed, respectively.

The five most well known potentials are:

Name	Symbol	Formula	Natural variables
Internal energy	U	$\int (TdS - pdV + \sum_i \mu_i dN_i)$	$S, V, \{N_i\}$
Helmholtz free energy	F	$U - TS$	$T, V, \{N_i\}$
Enthalpy	H	$U + pV$	$S, p, \{N_i\}$
Gibbs free energy	G	$U + pV - TS$	$T, p, \{N_i\}$
Landau Potential (Grand potential)	Ω, Φ_G	$U - TS - \sum_i \mu_i N_i$	$T, V, \{\mu_i\}$

where T is the temperature, S the entropy, p the pressure, V the volume, μ the chemical potential, N the number of particles in the system, and i is the count of particles types in the system.

Thermodynamic potentials can be derived from the energy balance equation applied to a thermodynamic system. Other thermodynamic potentials can also be obtained through Legendre transformation.

Applied Fields

- Atmospheric thermodynamics
- Biological thermodynamics
- Black hole thermodynamics
- Chemical thermodynamics
- Classical thermodynamics
- Equilibrium thermodynamics

- Industrial ecology (re: Exergy)

- Maximum entropy thermodynamics

- Non-equilibrium thermodynamics

- Philosophy of thermal and statistical physics

- Psychrometrics

- Quantum thermodynamics

- Statistical thermodynamics

- Thermoeconomics

Enthalpy and Internal Energy

Specific internal energy of the mixture, $u_{mix} = \sum_i X_i u_i$

Mass specific internal energy of the mixture, $\hat{u}_{mix} = \sum_i Y_i \hat{u}_i$

Specific enthalpy of the mixture, $h_{mix} = \sum_i X_i h_i$

Mass specific enthalpy of the mixture, $\hat{h}_{mix} = \sum_i Y_i \hat{h}_i$

Enthalpy of a species, $h^0_{i,T}(T) = \underset{\substack{\text{Heat of formation} \\ (\text{due to bond energy})}}{h^0_{f,298.15}} + \int_{298.15}^{T} \underset{\substack{\text{Sensible enthalpy} \\ (\text{associated with temperature})}}{C_{p,i}} dT$

Internal Energy of a species, $u^0_{i,T}(T) = u^0_{f,298.15} + \int_{298.15}^{T} C_{v,i} dT$

Effect of Temperature on Heat Capacity

Specific heats, C_p and C_v are functions of temperature for both ideal and real gases

Variation in sp. heat with temperature is caused by

- Vibrational energy

- Rotational energy

Sp. heat of mono atomic gases does not vary with temperature, why?

Their internal energy is contributed by translational energy.

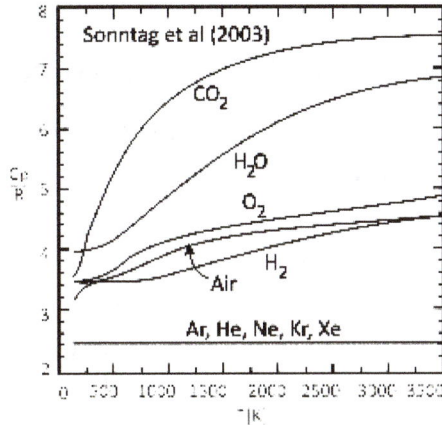

Sonntag et al (2003)

Laws of Thermodynamics

The four laws of thermodynamics define fundamental physical quantities (temperature, energy, and entropy) that characterize thermodynamic systems at thermal equilibrium. The laws describe how these quantities behave under various circumstances, and forbid certain phenomena (such as perpetual motion).

The four laws of thermodynamics are:

- Zeroth law of thermodynamics: If two systems are in thermal equilibrium with a third system, they are in thermal equilibrium with each other. This law helps define the notion of temperature.

- First law of thermodynamics: When energy passes, as work, as heat, or with matter, into or out from a system, the system's internal energy changes in accord with the law of conservation of energy. Equivalently, perpetual motion machines of the first kind are impossible.

- Second law of thermodynamics: In a natural thermodynamic process, the sum of the entropies of the interacting thermodynamic systems increases. Equivalently, perpetual motion machines of the second kind are impossible.

- Third law of thermodynamics: The entropy of a system approaches a constant value as the temperature approaches absolute zero. With the exception of non-crystalline solids (glasses) the entropy of a system at absolute zero is typically close to zero, and is equal to the logarithm of the product of the quantum ground states.

There have been suggestions of additional laws, but none of them achieves the generality of the four accepted laws, and they are not mentioned in standard textbooks.

The laws of thermodynamics are important fundamental laws in physics and they are applicable in other natural sciences.

Zeroth Law

The zeroth law of thermodynamics may be stated in the following form:

If two systems are both in thermal equilibrium with a third system then they are in thermal equilibrium with each other.

The law is intended to allow the existence of an empirical parameter, the temperature, as a property of a system such that systems in thermal equilibrium with each other have the same temperature. The law as stated here is compatible with the use of a particular physical body, for example a mass of gas, to match temperatures of other bodies, but does not justify regarding temperature as a quantity that can be measured on a scale of real numbers.

Though this version of the law is one of the more commonly stated, it is only one of a diversity of statements that are labeled as "the zeroth law" by competent writers. Some statements go further so as to supply the important physical fact that temperature is one-dimensional, that one can conceptually arrange bodies in real number sequence from colder to hotter. Perhaps there exists no unique "best possible statement" of the "zeroth law", because there is in the literature a range of formulations of the principles of thermodynamics, each of which call for their respectively appropriate versions of the law.

Although these concepts of temperature and of thermal equilibrium are fundamental to thermodynamics and were clearly stated in the nineteenth century, the desire to explicitly number the above law was not widely felt until Fowler and Guggenheim did so in the 1930s, long after the first, second, and third law were already widely understood and recognized. Hence it was numbered the zeroth law. The importance of the law as a foundation to the earlier laws is that it allows the definition of temperature in a non-circular way without reference to entropy, its conjugate variable. Such a temperature definition is said to be 'empirical'.

First Law

The first law of thermodynamics may be stated in several ways :

> The increase in internal energy of a closed system is equal to total of the energy added to the system. In particular, if the energy entering the system is supplied as heat and energy leaves the system as work, the heat is accounted as positive and the work is accounted as negative.

$$\Delta U_{system} = Q - W$$

> In the case of a thermodynamic cycle of a closed system, which returns to its original state, the heat Q_{in} supplied to the system in one stage of the cycle, minus

the heat Q_{out} removed from it in another stage of the cycle, plus the work added to the system W_{in} equals the work that leaves the system W_{out}.

$$\Delta U_{system(full\ cycle)} = 0$$

hence, for a full cycle,

$$Q = Q_{in} - Q_{out} + W_{in} - W_{out} = W_{net}$$

For the particular case of a thermally isolated system (adiabatically isolated), the change of the internal energy of an adiabatically isolated system can only be the result of the work added to the system, because the adiabatic assumption is: $Q = 0$.

$$\Delta U_{system} = U_{final} - U_{initial} = W_{in} - W_{out}$$

More specifically, the First Law encompasses several principles:

- The law of conservation of energy.

  This states that energy can be neither created nor destroyed. However, energy can change forms, and energy can flow from one place to another. A particular consequence of the law of conservation of energy is that the total energy of an isolated system does not change.

- The concept of internal energy and its relationship to temperature.

  If a system has a definite temperature, then its total energy has three distinguishable components. If the system is in motion as a whole, it has kinetic energy. If the system as a whole is in an externally imposed force field (e.g. gravity), it has potential energy relative to some reference point in space. Finally, it has internal energy, which is a fundamental quantity of thermodynamics. The establishment of the concept of internal energy distinguishes the first law of thermodynamics from the more general law of conservation of energy.

$$E_{total} = KE_{system} + PE_{system} + U_{system}$$

The internal energy of a substance can be explained as the sum of the diverse kinetic energies of the erratic microscopic motions of its constituent atoms, and of the potential energy of interactions between them. Those microscopic energy terms are collectively called the substance's internal energy (U), and are accounted for by macroscopic thermodynamic property. The total of the kinetic energies of microscopic motions of the constituent atoms increases as the system's temperature increases; this assumes no other interactions at the microscopic level of the system such as chemical reactions, potential energy of constituent atoms with respect to each other.

- Work is a process of transferring energy to or from a system in ways that can be described by macroscopic mechanical forces exerted by factors in the surroundings, outside the system. Examples are an externally driven shaft agitating a stirrer within the system, or an externally imposed electric field that polarizes the material of the system, or a piston that compresses the system. Unless otherwise stated, it is customary to treat work as occurring without its dissipation to the surroundings. Practically speaking, in all natural process, some of the work is dissipated by internal friction or viscosity. The work done by the system can come from its overall kinetic energy, from its overall potential energy, or from its internal energy.

For example, when a machine (not a part of the system) lifts a system upwards, some energy is transferred from the machine to the system. The system's energy increases as work is done on the system and in this particular case, the energy increase of the system is manifested as an increase in the system's gravitational potential energy. Work added to the system increases the Potential Energy of the system:

$$W = \Delta \mathrm{PE}_{system}$$

Or in general, the energy added to the system in the form of work can be partitioned to kinetic, potential or internal energy forms:

$$W = \Delta \mathrm{KE}_{system} + \Delta \mathrm{PE}_{system} + \Delta U_{system}$$

- When matter is transferred into a system, that masses' associated internal energy and potential energy are transferred with it.

$$\left(u \Delta M \right)_{in} = \Delta U_{system}$$

where u denotes the internal energy per unit mass of the transferred matter, as measured while in the surroundings; and ΔM denotes the amount of transferred mass.

- The flow of heat is a form of energy transfer.

Heating is a natural process of moving energy to or from a system other than by work or the transfer of matter. Direct passage of heat is only from a hotter to a colder system.

If the system has rigid walls that are impermeable to matter, and consequently energy cannot be transferred as work into or out from the system, and no external long-range force field affects it that could change its internal energy, then the internal energy can only be changed by the transfer of energy as heat:

$$\Delta U_{system} = Q$$

where Q denotes the amount of energy transferred into the system as heat.

Combining these principles leads to one traditional statement of the first law of thermodynamics: it is not possible to construct a machine which will perpetually output work without an equal amount of energy input to that machine. Or more briefly, a perpetual motion machine of the first kind is impossible.

Second Law

The second law of thermodynamics indicates the irreversibility of natural processes, and, in many cases, the tendency of natural processes to lead towards spatial homogeneity of matter and energy, and especially of temperature. It can be formulated in a variety of interesting and important ways.

It implies the existence of a quantity called the entropy of a thermodynamic system. In terms of this quantity it implies that

When two initially isolated systems in separate but nearby regions of space, each in thermodynamic equilibrium with itself but not necessarily with each other, are then allowed to interact, they will eventually reach a mutual thermodynamic equilibrium. The sum of the entropies of the initially isolated systems is less than or equal to the total entropy of the final combination. Equality occurs just when the two original systems have all their respective intensive variables (temperature, pressure) equal; then the final system also has the same values.

This statement of the second law is founded on the assumption, that in classical thermodynamics, the entropy of a system is defined only when it has reached internal thermodynamic equilibrium (thermodynamic equilibrium with itself).

The second law is applicable to a wide variety of processes, reversible and irreversible. All natural processes are irreversible. Reversible processes are a useful and convenient theoretical fiction, but do not occur in nature.

A prime example of irreversibility is in the transfer of heat by conduction or radiation. It was known long before the discovery of the notion of entropy that when two bodies initially of different temperatures come into thermal connection, then heat always flows from the hotter body to the colder one.

The second law tells also about kinds of irreversibility other than heat transfer, for example those of friction and viscosity, and those of chemical reactions. The notion of entropy is needed to provide that wider scope of the law.

According to the second law of thermodynamics, in a theoretical and fictive reversible heat transfer, an element of heat transferred, δQ, is the product of the temperature (T), both of the system and of the sources or destination of the heat, with the increment (dS) of the system's conjugate variable, its entropy (S)

$$\delta Q = T\, dS.$$

Entropy may also be viewed as a physical measure of the lack of physical information about the microscopic details of the motion and configuration of a system, when only the macroscopic states are known. This lack of information is often described as *disorder* on a microscopic or molecular scale. The law asserts that for two given macroscopically specified states of a system, there is a quantity called the difference of information entropy between them. This information entropy difference defines how much additional microscopic physical information is needed to specify one of the macroscopically specified states, given the macroscopic specification of the other - often a conveniently chosen reference state which may be presupposed to exist rather than explicitly stated. A final condition of a natural process always contains microscopically specifiable effects which are not fully and exactly predictable from the macroscopic specification of the initial condition of the process. This is why entropy increases in natural processes - the increase tells how much extra microscopic information is needed to distinguish the final macroscopically specified state from the initial macroscopically specified state.

Third Law

The third law of thermodynamics is sometimes stated as follows:

> *The entropy of a perfect crystal of any pure substance approaches zero as the temperature approaches absolute zero.*

At zero temperature the system must be in a state with the minimum thermal energy. This statement holds true if the perfect crystal has only one state with minimum energy. Entropy is related to the number of possible microstates according to:

$$S = k_B \ln \Omega$$

Where S is the entropy of the system, k_B Boltzmann's constant, and Ω the number of microstates (e.g. possible configurations of atoms). At absolute zero there is only 1 microstate possible (Ω=1 as all the atoms are identical for a pure substance and as a result all orders are identical as there is only one combination) and $\ln(1) = 0$.

A more general form of the third law that applies to a system such as a glass that may have more than one minimum microscopically distinct energy state, or may have a microscopically distinct state that is "frozen in" though not a strictly minimum energy state and not strictly speaking a state of thermodynamic equilibrium, at absolute zero temperature:

The entropy of a system approaches a constant value as the temperature approaches zero.

The constant value (not necessarily zero) is called the residual entropy of the system.

History

Circa 1797, Count Rumford (born Benjamin Thompson) showed that endless mechanical action can generate indefinitely large amounts of heat from a fixed amount of working substance thus challenging the caloric theory of heat, which held that there would be a finite amount of caloric heat/energy in a fixed amount of working substance. The first established thermodynamic principle, which eventually became the second law of thermodynamics, was formulated by Sadi Carnot in 1824. By 1860, as formalized in the works of those such as Rudolf Clausius and William Thomson, two established principles of thermodynamics had evolved, the first principle and the second principle, later restated as thermodynamic laws. By 1873, for example, thermodynamicist Josiah Willard Gibbs, in his memoir *Graphical Methods in the Thermodynamics of Fluids*, clearly stated the first two absolute laws of thermodynamics. Some textbooks throughout the 20th century have numbered the laws differently. In some fields removed from chemistry, the second law was considered to deal with the efficiency of heat engines only, whereas what was called the third law dealt with entropy increases. Directly defining zero points for entropy calculations was not considered to be a law. Gradually, this separation was combined into the second law and the modern third law was widely adopted.

Stoichiometry

Stoichiometry is the calculation of relative quantities of reactants and products in chemical reactions.

Stoichiometry is founded on the law of conservation of mass where the total mass of the reactants equals the total mass of the products leading to the insight that the relations among quantities of reactants and products typically form a ratio of positive integers. This means that if the amounts of the separate reactants are known, then the amount of the product can be calculated. Conversely, if one reactant has a known quantity and the quantity of product can be empirically determined, then the amount of the other reactants can also be calculated.

This is illustrated in the image here, where the balanced equation is:

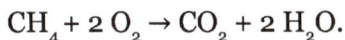

$$CH_4 + 2\,O_2 \rightarrow CO_2 + 2\,H_2O.$$

Here, one molecule of methane reacts with two molecules of oxygen gas to yield one molecule of carbon dioxide and two molecules of water. Stoichiometry measures these quantitative relationships, and is used to determine the amount of products/reactants that are produced/needed in a given reaction. Describing the quantitative relationships among substances as they participate in chemical reactions is known as *reaction stoichiometry*. In the example above, reaction stoichiometry measures the relationship between the methane and oxygen as they react to form carbon dioxide and water.

Because of the well known relationship of moles to atomic weights, the ratios that are arrived at by stoichiometry can be used to determine quantities by weight in a reaction described by a balanced equation. This is called *composition stoichiometry*.

Gas stoichiometry deals with reactions involving gases, where the gases are at a known temperature, pressure, and volume and can be assumed to be ideal gases. For gases, the volume ratio is ideally the same by the ideal gas law, but the mass ratio of a single reaction has to be calculated from the molecular masses of the reactants and products. In practice, due to the existence of isotopes, molar masses are used instead when calculating the mass ratio.

Etymology

The term *stoichiometry* was first used by Jeremias Benjamin Richter in 1792 when the first volume of Richter's *Stoichiometry or the Art of Measuring the Chemical Elements* was published. The term is derived from the Greek words *stoicheion* "element" and *metron* "measure". In patristic Greek, the word *Stoichiometria* was used by Nicephorus to refer to the number of line counts of the canonical New Testament and some of the Apocrypha.

Definition

A stoichiometric amount or stoichiometric ratio of a reagent is the optimum amount or ratio where, assuming that the reaction proceeds to completion:

1. All of the reagent is consumed

2. There is no deficiency of the reagent

3. There is no excess of the reagent.

Stoichiometry rests upon the very basic laws that help to understand it better, i.e., law of conservation of mass, the law of definite proportions (i.e., the law of constant composition), the law of multiple proportions and the law of reciprocal proportions. In general, chemical reactions combine in definite ratios of chemicals. Since chemical reactions can neither create nor destroy matter, nor transmute one element into another, the amount of each element must be the same throughout the overall reaction. For example, the number of atoms of a given element X on the reactant side must equal the number of atoms of that element on the product side, whether or not all of those atoms are actually involved in a reaction.

Chemical reactions, as macroscopic unit operations, consist of simply a very large number of elementary reactions, where a single molecule reacts with another molecule. As the reacting molecules (or moieties) consist of a definite set of atoms in an integer ratio, the ratio between reactants in a complete reaction is also in integer ratio. A reaction may consume more than one molecule, and the stoichiometric number counts this number, defined as positive for products (added) and negative for reactants (removed).

Different elements have a different atomic mass, and as collections of single atoms, molecules have a definite molar mass, measured with the unit mole (6.02×10^{23} individual molecules, Avogadro's constant). By definition, carbon-12 has a molar mass of 12 g/mol. Thus, to calculate the stoichiometry by mass, the number of molecules required for each reactant is expressed in moles and multiplied by the molar mass of each to give the mass of each reactant per mole of reaction. The mass ratios can be calculated by dividing each by the total in the whole reaction.

Elements in their natural state are mixtures of isotopes of differing mass, thus atomic masses and thus molar masses are not exactly integers. For instance, instead of an exact 14:3 proportion, 17.04 kg of ammonia consists of 14.01 kg of nitrogen and 3×1.01 kg of hydrogen, because natural nitrogen includes a small amount of nitrogen-15, and natural hydrogen includes hydrogen-2 (deuterium).

A stoichiometric reactant is a reactant that is consumed in a reaction, as opposed to a catalytic reactant, which is not consumed in the overall reaction because it reacts in one step and is regenerated in another step.

Converting Grams to Moles

Stoichiometry is not only used to balance chemical equations but also used in conversions, i.e., converting from grams to moles using molar mass as the conversion factor, or from grams to milliliters using density. For example, to find the amount of NaCl (sodium chloride) in 2.00 g, one would do the following:

$$\frac{2.00 \text{ g NaCl}}{58.44 \text{ g NaCl mol}^{-1}} = 0.034 \text{ mol}$$

In the above example, when written out in fraction form, the units of grams form a multiplicative identity, which is equivalent to one (g/g = 1), with the resulting amount in moles (the unit that was needed), as shown in the following equation,

$$\left(\frac{2.00 \text{ g NaCl}}{1} \right) \left(\frac{1 \text{ mol NaCl}}{58.44 \text{ g NaCl}} \right) = 0.034 \text{ mol}$$

Molar Proportion

Stoichiometry is often used to balance chemical equations (reaction stoichiometry). For example, the two diatomic gases, hydrogen and oxygen, can combine to form a liquid, water, in an exothermic reaction, as described by the following equation:

$$2\ H_2 + O_2 \rightarrow 2\ H_2O$$

Reaction stoichiometry describes the 2:1:2 ratio of hydrogen, oxygen, and water molecules in the above equation.

The molar ratio allows for conversion between moles of one substance and moles of another. For example, in the reaction

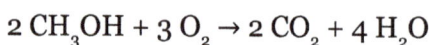

$$2\ CH_3OH + 3\ O_2 \rightarrow 2\ CO_2 + 4\ H_2O$$

the amount of water that will be produced by the combustion of 0.27 moles of CH_3OH is obtained using the molar ratio between CH_3OH and H_2O of 2 to 4.

$$\left(\frac{0.27\ \text{mol } CH_3OH}{1}\right)\left(\frac{4\ \text{mol } H_2O}{2\ \text{mol } CH_3OH}\right) = 0.54\ \text{mol } H_2O$$

The term stoichiometry is also often used for the molar proportions of elements in stoichiometric compounds (composition stoichiometry). For example, the stoichiometry of hydrogen and oxygen in H_2O is 2:1. In stoichiometric compounds, the molar proportions are whole numbers.

Determining Amount of Product

Stoichiometry can also be used to find the quantity of a product yielded by a reaction. If a piece of solid copper (Cu) were added to an aqueous solution of silver nitrate ($AgNO_3$), the silver (Ag) would be replaced in a single displacement reaction forming aqueous copper(II) nitrate ($Cu(NO_3)_2$) and solid silver. How much silver is produced if 16.00 grams of Cu is added to the solution of excess silver nitrate?

The following steps would be used:

1. Write and balance the equation

2. Mass to moles: Convert grams of Cu to moles of Cu

3. Mole ratio: Convert moles of Cu to moles of Ag produced

4. Mole to mass: Convert moles of Ag to grams of Ag produced

The complete balanced equation would be:

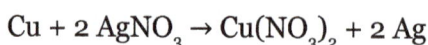

$$Cu + 2\ AgNO_3 \rightarrow Cu(NO_3)_2 + 2\ Ag$$

For the mass to mole step, the mass of copper (16.00 g) would be converted to moles of copper by dividing the mass of copper by its molecular mass: 63.55 g/mol.

$$\left(\frac{16.00 \text{ g Cu}}{1}\right)\left(\frac{1 \text{ mol Cu}}{63.55 \text{ g Cu}}\right) = 0.2518 \text{ mol Cu}$$

Now that the amount of Cu in moles (0.2518) is found, we can set up the mole ratio. This is found by looking at the coefficients in the balanced equation: Cu and Ag are in a 1:2 ratio.

$$\left(\frac{0.2518 \text{ mol Cu}}{1}\right)\left(\frac{2 \text{ mol Ag}}{1 \text{ mol Cu}}\right) = 0.5036 \text{ mol Ag}$$

Now that the moles of Ag produced is known to be 0.5036 mol, we convert this amount to grams of Ag produced to come to the final answer:

$$\left(\frac{0.5036 \text{ mol Ag}}{1}\right)\left(\frac{107.87 \text{ g Ag}}{1 \text{ mol Ag}}\right) = 54.32 \text{ g Ag}$$

This set of calculations can be further condensed into a single step:

$$m_{Ag} = \left(\frac{16.00 \text{ g Cu}}{1}\right)\left(\frac{1 \text{ mol Cu}}{63.55 \text{ g Cu}}\right)\left(\frac{2 \text{ mol Ag}}{1 \text{ mol Cu}}\right)\left(\frac{107.87 \text{ g Ag}}{1 \text{ mol Ag}}\right) = 54.32 \text{ g}$$

Further Examples

For propane (C_3H_8) reacting with oxygen gas (O_2), the balanced chemical equation is:

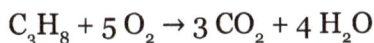

$$C_3H_8 + 5 O_2 \rightarrow 3 CO_2 + 4 H_2O$$

The mass of water formed if 120 g of propane (C_3H_8) is burned in excess oxygen is then

$$mH_2O = \left(\frac{120. \text{ g } C_3H_8}{1}\right)\left(\frac{1 \text{ mol } C_3H_8}{44.09 \text{ g } C_3H_8}\right)\left(\frac{4 \text{ mol } H_2O}{1 \text{ mol } C_3H_8}\right)\left(\frac{18.02 \text{ g } H_2O}{1 \text{ mol } H_2O}\right) = 196 \text{ g}$$

Stoichiometric Ratio

Stoichiometry is also used to find the right amount of one reactant to "completely" react with the other reactant in a chemical reaction – that is, the stoichiometric amounts that would result in no leftover reactants when the reaction takes place. An example is shown below using the thermite reaction,

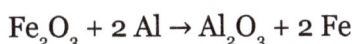

$$Fe_2O_3 + 2 Al \rightarrow Al_2O_3 + 2 Fe$$

This equation shows that 1 mole of iron(III) oxide and 2 moles of aluminum will produce 1 mole of aluminium oxide and 2 moles of iron. So, to completely react with 85.0 g of iron(III) oxide (0.532 mol), 28.7 g (1.06 mol) of aluminium are needed.

$$m_{Al} = \left(\frac{85.0 \text{ g Fe}_2\text{O}_3}{1} \right)\left(\frac{1 \text{ mol Fe}_2\text{O}_3}{159.7 \text{ g Fe}_2\text{O}_3} \right)\left(\frac{2 \text{ mol Al}}{1 \text{ mol Fe}_2\text{O}_3} \right)\left(\frac{26.98 \text{ g Al}}{1 \text{ mol Al}} \right) = 28.7 \text{ g}$$

Limiting Reagent and Percent Yield

The limiting reagent is the reagent that limits the amount of product that can be formed and is completely consumed when the reaction is complete. An excess reactant is a reactant that is left over once the reaction has stopped due to the limiting reactant being exhausted.

Consider the equation of roasting lead(II) sulfide (PbS) in oxygen (O_2) to produce lead(II) oxide (PbO) and sulfur dioxide (SO_2):

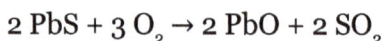

$$2 \text{ PbS} + 3 \text{ O}_2 \rightarrow 2 \text{ PbO} + 2 \text{ SO}_2$$

To determine the theoretical yield of lead(II) oxide if 200.0 g of lead(II) sulfide and 200.0 g of oxygen are heated in an open container:

$$m_{PbO} = \left(\frac{200.0 \text{ g PbS}}{1} \right)\left(\frac{1 \text{ mol PbS}}{239.27 \text{ g PbS}} \right)\left(\frac{2 \text{ mol PbO}}{2 \text{ mol PbS}} \right)\left(\frac{223.2 \text{ g PbO}}{1 \text{ mol PbO}} \right) = 186.6 \text{ g}$$

$$m_{PbO} = \left(\frac{200.0 \text{ g O}_2}{1} \right)\left(\frac{1 \text{ mol O}_2}{32.00 \text{ g O}_2} \right)\left(\frac{2 \text{ mol PbO}}{3 \text{ mol O}_2} \right)\left(\frac{223.2 \text{ g PbO}}{1 \text{ mol PbO}} \right) = 930.0 \text{ g}$$

Because a lesser amount of PbO is produced for the 200.0 g of PbS, it is clear that PbS is the limiting reagent.

In reality, the actual yield is not the same as the stoichiometrically-calculated theoretical yield. Percent yield, then, is expressed in the following equation:

$$\text{percent yield} = \frac{\text{actual yield}}{\text{theoretical yield}}$$

If 170.0 g of lead(II) oxide is obtained, then the percent yield would be calculated as follows:

$$\text{percent yield} = \frac{170.0 \text{ g PbO}}{186.6 \text{ g PbO}} = 91.12\%$$

Example

Consider the following reaction, in which iron(III) chloride reacts with hydrogen sulfide to produce iron(III) sulfide and hydrogen chloride:

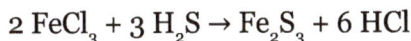

$$2\ FeCl_3 + 3\ H_2S \rightarrow Fe_2S_3 + 6\ HCl$$

Suppose 90.0 g of $FeCl_3$ reacts with 52.0 g of H_2S. To find the limiting reagent and the mass of HCl produced by the reaction, we could set up the following equations:

$$m_{HCl} = \left(\frac{90.0\ g\ FeCl_3}{1}\right)\left(\frac{1\ mol\ FeCl_3}{162\ g\ FeCl_3}\right)\left(\frac{6\ mol\ HCl}{2\ mol\ FeCl_3}\right)\left(\frac{36.5\ g\ HCl}{1\ mol\ HCl}\right) = 60.8\ g$$

$$m_{HCl} = \left(\frac{52.0\ g\ H_2S}{1}\right)\left(\frac{1\ mol\ H_2S}{34.1\ g\ H_2S}\right)\left(\frac{6\ mol\ HCl}{3\ mol\ H_2S}\right)\left(\frac{36.5\ g\ HCl}{1\ mol\ HCl}\right) = 111\ g$$

Thus, the limiting reagent is $FeCl_3$ and the amount of HCl produced is 60.8 g.

To find what mass of excess reagent (H_2S) remains after the reaction, we would set up the calculation to find out how much H_2S reacts completely with the 90.0 g $FeCl_3$:

$$m_{H_2S} = \left(\frac{90.0\ g\ FeCl_3}{1}\right)\left(\frac{1\ mol\ FeCl_3}{162\ g\ FeCl_3}\right)\left(\frac{3\ mol\ H_2S}{2\ mol\ FeCl_3}\right)\left(\frac{34.1\ g\ H_2S}{1\ mol\ H_2S}\right) = 28.4\ g\ reacted$$

By subtracting this amount from the original amount of H_2S, we can come to the answer:

$$52.0\ g\ H_2S - 28.4\ g\ H_2S = 23.6\ g\ H_2S\ excess$$

Different Stoichiometries in Competing Reactions

Often, more than one reaction is possible given the same starting materials. The reactions may differ in their stoichiometry. For example, the methylation of benzene (C_6H_6), through a Friedel–Crafts reaction using $AlCl_3$ as a catalyst, may produce singly methylated ($C_6H_5CH_3$), doubly methylated ($C_6H_4(CH_3)_2$), or still more highly methylated ($C_6H_{6-n}(CH_3)_n$) products, as shown in the following example,

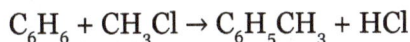

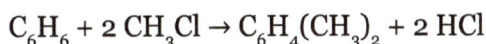

$$C_6H_6 + CH_3Cl \rightarrow C_6H_5CH_3 + HCl$$

$$C_6H_6 + 2\ CH_3Cl \rightarrow C_6H_4(CH_3)_2 + 2\ HCl$$

$$C_6H_6 + n\ CH_3Cl \rightarrow C_6H_{6-n}(CH_3)_n + n\ HCl$$

In this example, which reaction takes place is controlled in part by the relative concentrations of the reactants.

Stoichiometric Coefficient

In lay terms, the *stoichiometric coefficient* (or *stoichiometric number* in the IUPAC nomenclature) of any given component is the number of molecules that participate in the reaction as written.

For example, in the reaction $CH_4 + 2 O_2 \rightarrow CO_2 + 2 H_2O$, the stoichiometric coefficient of CH_4 is −1, the stoichiometric coefficient of O_2 is −2, for CO_2 it would be +1 and for H_2O it is +2.

In more technically precise terms, the stoichiometric coefficient in a chemical reaction system of the ith component is defined as

$$\nu_i = \frac{\Delta N_i}{\Delta \xi}$$

or

$$\Delta N_i = \nu_i \Delta \xi$$

where N_i is the number of molecules of i, and ξ is the progress variable or extent of reaction.

The extent of reaction ξ can be regarded as [the amount of] a real (or hypothetical) product, one molecule of which produced each time the reaction event occurs. It is the extensive quantity describing the progress of a chemical reaction equal to the number of chemical transformations, as indicated by the reaction equation on a molecular scale, divided by the Avogadro constant (in essence, it is the amount of chemical transformations). The change in the extent of reaction is given by $d\xi = dn_B/\nu_B$, where ν_B is the stoichiometric number of any reaction entity B (reactant or product) and n_B is the corresponding amount.

The stoichiometric coefficient ν_i represents the degree to which a chemical species participates in a reaction. The convention is to assign negative coefficients to *reactants* (which are consumed) and positive ones to *products*. However, any reaction may be viewed as going in the reverse direction, and all the coefficients then change sign (as does the free energy). Whether a reaction actually *will* go in the arbitrarily selected forward direction or not depends on the amounts of the substances present at any given time, which determines the kinetics and thermodynamics, i.e., whether equilibrium lies to the *right* or the *left*.

In reaction mechanisms, stoichiometric coefficients for each step are always integers, since elementary reactions always involve whole molecules. If one uses a composite representation of an overall reaction, some may be rational fractions. There are often chemical species present that do not participate in a reaction; their stoichiometric coefficients are therefore zero. Any chemical species that is regenerated, such as a catalyst, also has a stoichiometric coefficient of zero.

The simplest possible case is an isomerization

$$A \to B$$

in which $v_B = 1$ since one molecule of B is produced each time the reaction occurs, while $v_A = -1$ since one molecule of A is necessarily consumed. In any chemical reaction, not only is the total mass conserved but also the numbers of atoms of each kind are conserved, and this imposes corresponding constraints on possible values for the stoichiometric coefficients.

There are usually multiple reactions proceeding simultaneously in any natural reaction system, including those in biology. Since any chemical component can participate in several reactions simultaneously, the stoichiometric coefficient of the ith component in the kth reaction is defined as

$$v_{ik} = \frac{\partial N_i}{\partial \xi_k}$$

so that the total (differential) change in the amount of the ith component is

$$dN_i = \sum_k v_{ik} d\xi_k.$$

Extents of reaction provide the clearest and most explicit way of representing compositional change, although they are not yet widely used.

With complex reaction systems, it is often useful to consider both the representation of a reaction system in terms of the amounts of the chemicals present $\{ N_i \}$ (state variables), and the representation in terms of the actual compositional degrees of freedom, as expressed by the extents of reaction $\{ \xi_k \}$. The transformation from a vector expressing the extents to a vector expressing the amounts uses a rectangular matrix whose elements are the stoichiometric coefficients $[v_{ik}]$.

The maximum and minimum for any ξ_k occur whenever the first of the reactants is depleted for the forward reaction; or the first of the "products" is depleted if the reaction as viewed as being pushed in the reverse direction. This is a purely kinematic restriction on the reaction simplex, a hyperplane in composition space, or Nspace, whose dimensionality equals the number of *linearly-independent* chemical reactions. This is necessarily less than the number of chemical components, since each reaction manifests a relation between at least two chemicals. The accessible region of the hyperplane depends on the amounts of each chemical species actually present, a contingent fact. Different such amounts can even generate different hyperplanes, all sharing the same algebraic stoichiometry.

In accord with the principles of chemical kinetics and thermodynamic equilibrium, every chemical reaction is *reversible*, at least to some degree, so that each equilibrium point must be an interior point of the simplex. As a consequence, extrema for the ξs will not occur unless an experimental system is prepared with zero initial amounts of some products.

The number of *physically*-independent reactions can be even greater than the number of chemical components, and depends on the various reaction mechanisms. For example, there may be two (or more) reaction *paths* for the isomerism above. The reaction may occur by itself, but faster and with different intermediates, in the presence of a catalyst.

The (dimensionless) "units" may be taken to be molecules or moles. Moles are most commonly used, but it is more suggestive to picture incremental chemical reactions in terms of molecules. The Ns and ξs are reduced to molar units by dividing by Avogadro's number. While dimensional mass units may be used, the comments about integers are then no longer applicable.

Stoichiometry Matrix

In complex reactions, stoichiometries are often represented in a more compact form called the stoichiometry matrix. The stoichiometry matrix is denoted by the symbol N.

If a reaction network has n reactions and m participating molecular species then the stoichiometry matrix will have correspondingly m rows and n columns.

For example, consider the system of reactions shown below:

$$S_1 \rightarrow S_2$$

$$5\,S_3 + S_2 \rightarrow 4\,S_3 + 2\,S_2$$

$$S_3 \rightarrow S_4$$

$$S_4 \rightarrow S_5$$

This systems comprises four reactions and five different molecular species. The stoichiometry matrix for this system can be written as:

$$\mathbf{N} = \begin{bmatrix} -1 & 0 & 0 & 0 \\ 1 & 1 & 0 & 0 \\ 0 & -1 & -1 & 0 \\ 0 & 0 & 1 & -1 \\ 0 & 0 & 0 & 1 \end{bmatrix}$$

where the rows correspond to S_1, S_2, S_3, S_4 and S_5, respectively. Note that the process of converting a reaction scheme into a stoichiometry matrix can be a lossy transformation, for example, the stoichiometries in the second reaction simplify when included in the matrix. This means that it is not always possible to recover the original reaction scheme from a stoichiometry matrix.

Often the stoichiometry matrix is combined with the rate vector, v, and the species vector, S to form a compact equation describing the rates of change of the molecular species:

$$\frac{d\mathbf{S}}{dt} = \mathbf{N} \cdot \mathbf{v}.$$

Gas Stoichiometry

Gas stoichiometry is the quantitative relationship (ratio) between reactants and products in a chemical reaction with reactions that produce gases. Gas stoichiometry applies when the gases produced are assumed to be ideal, and the temperature, pressure, and volume of the gases are all known. The ideal gas law is used for these calculations. Often, but not always, the standard temperature and pressure (STP) are taken as 0°C and 1 bar and used as the conditions for gas stoichiometric calculations.

Gas stoichiometry calculations solve for the unknown volume or mass of a gaseous product or reactant. For example, if we wanted to calculate the volume of gaseous NO_2 produced from the combustion of 100 g of NH_3, by the reaction:

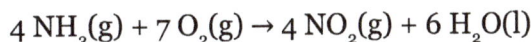

$$4\ NH_3(g) + 7\ O_2(g) \rightarrow 4\ NO_2(g) + 6\ H_2O(l)$$

we would carry out the following calculations:

$$100 g\,NH_3 \cdot \frac{1 mol\,NH_3}{17.034 g\,NH_3} = 5.871 mol\,NH_3$$

There is a 1:1 molar ratio of NH_3 to NO_2 in the above balanced combustion reaction, so 5.871 mol of NO_2 will be formed. We will employ the ideal gas law to solve for the volume at 0°C (273.15 K) and 1 atmosphere using the gas law constant of $R = 0.08206$ L·atm·K^{-1}·mol^{-1} :

$$PV = nRT$$
$$V = \frac{nRT}{P}$$
$$= \frac{5.871 \cdot 0.08206 \cdot 273.15}{1}$$
$$= 131.597 L\,NO_2$$

Gas stoichiometry often involves having to know the molar mass of a gas, given the density of that gas. The ideal gas law can be re-arranged to obtain a relation between the density and the molar mass of an ideal gas:

$$\rho = \frac{m}{V} \quad \text{and} \quad n = \frac{m}{M}$$

and thus:

$$\rho = \frac{MP}{RT}$$

where:

- P = absolute gas pressure

- V = gas volume

- n = amount (measured in moles)

- R = universal ideal gas law constant

- T = absolute gas temperature

- ρ = gas density at T and P

- m = mass of gas

- M = molar mass of gas

Stoichiometric Air-to-fuel Ratios of Common Fuels

In the combustion reaction, oxygen reacts with the fuel, and the point where exactly all oxygen is consumed and all fuel burned is defined as the stoichiometric point. With more oxygen (overstoichiometric combustion), some of it stays unreacted. Likewise, if the combustion is incomplete due to lack of sufficient oxygen, fuel remains unreacted. (Unreacted fuel may also remain because of slow combustion or insufficient mixing of fuel and oxygen – this is not due to stoichiometry). Different hydrocarbon fuels have different contents of carbon, hydrogen and other elements, thus their stoichiometry varies.

Fuel	Ratio by mass	Ratio by volume	Percent fuel by mass
Gasoline	14.7 : 1	—	6.8%
Natural gas	17.2 : 1	9.7 : 1	5.8%
Propane (LP)	15.67 : 1	23.9 : 1	6.45%
Ethanol	9 : 1	—	11.1%
Methanol	6.47 : 1	—	15.6%
n-Butanol	11.2 : 1	—	8.2%
Hydrogen	34.3 : 1	2.39 : 1	2.9%
Diesel	14.5 : 1	—	6.8%
Methane	17.19 : 1	9.52 : 1	5.5%

Gasoline engines can run at stoichiometric air-to-fuel ratio, because gasoline is quite volatile and is mixed (sprayed or carburetted) with the air prior to ignition. Diesel engines, in contrast, run lean, with more air available than simple stoichiometry would require. Diesel fuel is less volatile and is effectively burned as it is injected, leaving less time for evaporation and mixing. Thus, it would form soot (black smoke) at stoichiometric ratio.

Conservation of Mass

Russian scientist Mikhail Lomonosov discovered the law of mass conservation in 1756 by experiments, and came to the conclusion that phlogiston theory is incorrect.

Antoine Lavoisier's discovery of the Law of Conservation of Mass led to many new findings in the 19th century. Joseph Proust's law of definite proportions and John Dalton's atomic theory branched from the discoveries of Antoine Lavoisier. Lavoisier's quantitative experiments revealed that combustion involved oxygen rather than what was previously thought to be phlogiston.

The law of conservation of mass or principle of mass conservation states that for any system closed to all transfers of matter and energy, the mass of the system must remain constant over time, as system mass cannot change quantity if it is not added or removed. Hence, the quantity of mass is "conserved" over time. The law implies that mass can neither be created nor destroyed, although it may be rearranged in space, or the entities associated with it may be changed in form, as for example when light or physical work is transformed into particles that contribute the same mass to the system as the light or work had contributed. Thus, during any chemical reaction, nuclear reaction, or radioactive decay in an isolated system, the total mass of the reactants or starting materials must be equal to the mass of the products.

The concept of mass conservation is widely used in many fields such as chemistry, mechanics, and fluid dynamics. Historically, mass conservation was discovered in chemical reactions by Antoine Lavoisier in the late 18th century, and was of crucial importance in the progress from alchemy to the modern natural science of chemistry.

The closely related concept of matter conservation was found to hold good in chemistry to such high approximation that it failed only for the high energies treated by the later refinements of relativity theory, but otherwise remains useful and sufficiently accurate for most chemical calculations, even in modern practice.

In special relativity, needed for accuracy when large energy transfers between systems is involved, the difference between thermodynamically closed and isolated systems becomes important, since conservation of mass is strictly and perfectly upheld only for so-called isolated systems, i.e. those completely isolated from all exchanges with the environment. In this circumstance, the mass–energy equivalence theorem states that mass conservation is equivalent to total energy conservation, which is the first law of thermodynamics. By contrast, for a thermodynamically closed system (i.e., one which is closed to exchanges of matter, but open to exchanges of non-material energy, such as heat and work, with the surroundings) mass is (usually) only approximately conserved. The input or output of non-material energy must change the mass of the system in relativity theory, although the change is usually small, since relatively large amounts of such energy (by comparison with ordinary experience) carry only a small amount of mass (again by ordinary standards of measurement).

In special relativity, mass is not *converted* to energy, since mass and energy cannot be destroyed, and energy in all of its forms always retains its equivalent amount of mass throughout any transformation to a different type of energy within a system (or translocation into or out of a system). Certain types of *matter* (a different concept) may be created or destroyed, but in all of these processes, the energy and mass associated with such matter remains unchanged in quantity (although type of energy associated with the matter may change form).

In general relativity, mass (and energy) conservation in expanding volumes of space is a complex concept, subject to different definitions, and neither mass nor energy is as strictly and simply conserved as is the case in special relativity and in Minkowski space.

History

An important idea in ancient Greek philosophy was that "Nothing comes from nothing", so that what exists now has always existed: no new matter can come into existence where there was none before. An explicit statement of this, along with the further principle that nothing can pass away into nothing, is found in Empedocles (approx. 490–430 BC): "For it is impossible for anything to come to be from what is not, and it cannot be brought about or heard of that what is should be utterly destroyed."

A further principle of conservation was stated by Epicurus (341–270 BC) who, describing the nature of the Universe, wrote that "the totality of things was always such as it is now, and always will be".

Jain philosophy, a non-creationist philosophy based on the teachings of Mahavira (6th century BC), states that the universe and its constituents such as matter cannot be destroyed or created. The Jain text Tattvarthasutra (2nd century AD) states that a substance is permanent, but its modes are characterised by creation and destruction. A principle of the conservation of matter was also stated by Nasīr al-Dīn al-Tūsī (1201–1274). He wrote that "A body of matter cannot disappear completely. It only changes its form, condition, composition, color and other properties and turns into a different complex or elementary matter".

Mass Conservation in Chemistry

The principle of conservation of mass was first outlined by Mikhail Lomonosov (1711–1765) in 1748. He proved it by experiments—though this is sometimes challenged. Antoine Lavoisier (1743–1794) had expressed these ideas in 1774. Others whose ideas pre-dated the work of Lavoisier include Joseph Black (1728–1799), Henry Cavendish (1731–1810), and Jean Rey (1583–1645).

The conservation of mass was obscure for millennia because of the buoyancy effect of the Earth's atmosphere on the weight of gases. For example, a piece of wood weighs less after burning; this seemed to suggest that some of its mass disappears, or is transformed or lost. This was not disproved until careful experiments were performed in which chemical reactions such as rusting were allowed to take place in sealed glass ampoules; it was found that the chemical reaction did not change the weight of the sealed container and its contents. The vacuum pump also enabled the weighing of gases using scales.

Once understood, the conservation of mass was of great importance in progressing from alchemy to modern chemistry. Once early chemists realized that chemical substances never disappeared but were only transformed into other substances with the same weight, these scientists could for the first time embark on quantitative studies of the transformations of substances. The idea of mass conservation plus a surmise that certain "elemental substances" also could not be transformed into others by chemical reactions, in turn led to an understanding of chemical elements, as well as the idea that all chemical processes and transformations (such as burning and metabolic reactions) are reactions between invariant amounts or weights of these chemical elements.

Following the pioneering work of Lavoisier the prolonged and exhaustive experiments of Jean Stas supported the strict accuracy of this law in chemical reactions, even though they were carried out with other intentions. His research indicated that in certain reactions the loss or gain could not have been more than from 2 to 4 parts in 100,000. The difference in the accuracy aimed at and attained by Lavoisier on the one hand, and by Morley and Stas on the other, is enormous.

Generalization

In special relativity, the conservation of mass does not apply if the system is open and energy escapes. However, it does continue to apply to totally closed (isolated) systems. If energy cannot escape a system, its mass cannot decrease. In relativity theory, so long as any type of energy is retained within a system, this energy exhibits mass.

Also, mass must be differentiated from matter, since matter may *not* be perfectly conserved in isolated systems, even though mass is always conserved in such systems. However, matter is so nearly conserved in chemistry that violations of matter conservation were not measured until the nuclear age, and the assumption of matter conservation remains an important practical concept in most systems in chemistry and other studies that do not involve the high energies typical of radioactivity and nuclear reactions.

The Mass Associated with Chemical Amounts of Energy is too Small to Measure

The change in mass of certain kinds of open systems where atoms or massive particles are not allowed to escape, but other types of energy (such as light or heat) are allowed to enter or escape, went unnoticed during the 19th century, because the change in mass associated with addition or loss of small quantities of thermal or radiant energy in chemical reactions is very small. (In theory, mass would not change at all for experiments conducted in isolated systems where heat and work were not allowed in or out.)

The theoretical association of all energy with mass was made by Albert Einstein in 1905. However Max Planck pointed out that the change in mass of systems as a result of extraction or addition of chemical energy, as predicted by Einstein's theory, is so small that it could not be measured with available instruments, for example as a test of Einstein's theory. Einstein speculated that the energies associated with newly discovered radioactivity were significant enough, compared with the mass of systems producing them, to enable their mass-change to be measured, once the energy of the reaction had been removed from the system. This later indeed proved to be possible, although it was eventually to be the first artificial nuclear transmutation reaction in 1932, demonstrated by Cockcroft and Walton, that proved the first successful test of Einstein's theory regarding mass-loss with energy-loss.

Mass Conservation Remains Correct if Energy is not Lost

The conservation of relativistic mass implies the viewpoint of a single observer (or the view from a single inertial frame) since changing inertial frames may result in a change of the total energy (relativistic energy) for systems, and this quantity determines the relativistic mass.

The principle that the mass of a system of particles must be equal to the sum of their rest masses, even though true in classical physics, may be false in special relativity. The reason that rest masses cannot be simply added is that this does not take into account other forms of energy, such as kinetic and potential energy, and massless particles such as photons, all of which may (or may not) affect the total mass of systems.

For moving massive particles in a system, examining the rest masses of the various particles also amounts to introducing many different inertial observation frames (which is prohibited if total system energy and momentum are to be conserved), and also when in the rest frame of one particle, this procedure ignores the momenta of other particles, which affect the system mass if the other particles are in motion in this frame.

For the special type of mass called invariant mass, changing the inertial frame of observation for a whole closed system has no effect on the measure of invariant mass of the system, which remains both conserved and invariant (unchanging), even for different observers who view the entire system. Invariant mass is a system combination of energy and momentum, which is invariant for any observer, because in any inertial frame, the energies and momenta of the various particles always add to the same quantity (the momentum may be negative, so the addition amounts to a subtraction). The invariant mass is the relativistic mass of the system when viewed in the center of momentum frame. It is the minimum mass which a system may exhibit, as viewed from all possible inertial frames.

The conservation of both relativistic and invariant mass applies even to systems of particles created by pair production, where energy for new particles may come from kinetic energy of other particles, or from one or more photons as part of a system that includes other particles besides a photon. Again, neither the relativistic nor the invariant mass of totally closed (that is, isolated) systems changes when new particles are created. However, different inertial observers will disagree on the value of this conserved mass, if it is the relativistic mass (i.e., relativistic mass is conserved by not invariant). However, all observers agree on the value of the conserved mass if the mass being measured is the invariant mass (i.e., invariant mass is both conserved and invariant).

The mass-energy equivalence formula gives a different prediction in non-isolated systems, since if energy is allowed to escape a system, both relativistic mass and invariant mass will escape also. In this case, the mass-energy equivalence formula predicts that the *change* in mass of a system is associated with the *change* in its energy due to energy being added or subtracted: $\Delta m = \Delta E / c^2$. This form involving changes was the form in which this famous equation was originally presented by Einstein. In this sense, mass changes in any system are explained simply if the mass of the energy added or removed from the system, are taken into account.

The formula implies that bound systems have an invariant mass (rest mass for the system) less than the sum of their parts, if the binding energy has been allowed to escape the system after the system has been bound. This may happen by converting system potential energy into some other kind of active energy, such as kinetic energy or photons, which easily escape a bound system. The difference in system masses, called a mass defect, is a measure of the binding energy in bound systems – in other words, the energy needed to break the system apart. The greater the mass defect, the larger the binding energy. The binding energy (which itself has mass) must be released (as light or heat) when the parts combine to form the bound system, and this is the reason the mass of the bound system decreases when the energy leaves the system. The total invariant mass is actually conserved, when the mass of the binding energy that has escaped, is taken into account.

Exceptions or Caveats to Mass/Matter Conservation

Matter is not Perfectly Conserved

The principle of *matter* conservation may be considered as an approximate physical law that is true only in the classical sense, without consideration of special relativity and quantum mechanics. It is approximately true except in certain high energy applications.

A particular difficulty with the idea of conservation of "matter" is that "matter" is not a well-defined word scientifically, and when particles that are considered to be "matter" (such as electrons and positrons) are annihilated to make photons (which are often *not* considered matter) then conservation of matter does not take place over time, even within isolated systems. However, matter is conserved to such an extent that matter conservation may be safely assumed in chemical reactions and all situations in which radioactivity and nuclear reactions are not involved.

Even when matter is not conserved, the collection of mass and energy within the system are conserved.

Open Systems and Thermodynamically Closed Systems

Mass is also not generally conserved in *open* systems. Such is the case when various forms of energy are allowed into, or out of, the system. However, again unless radioactivity or nuclear reactions are involved, the amount of energy escaping such systems as heat, work, or electromagnetic radiation is usually too small to be measured as a decrease in system mass.

The law of mass conservation for isolated systems (totally closed to all mass and energy), as viewed over time from any single inertial frame, continues to be true in modern physics. The reason for this is that relativistic equations show that even "massless" particles such as photons still add mass and energy to isolated systems, allowing mass (though not matter) to be strictly conserved in all processes where energy does not escape the system. In relativity, different observers may disagree as to the particular *value* of the conserved mass of a given system, but each observer will agree that this value does not change over time as long as the system is isolated (totally closed to everything).

General Relativity

In general relativity, the total invariant mass of photons in an expanding volume of space will decrease, due to the red shift of such an expansion. The conservation of both mass and energy therefore depends on various corrections made to energy in the theory, due to the changing gravitational potential energy of such systems.

Thermochemistry

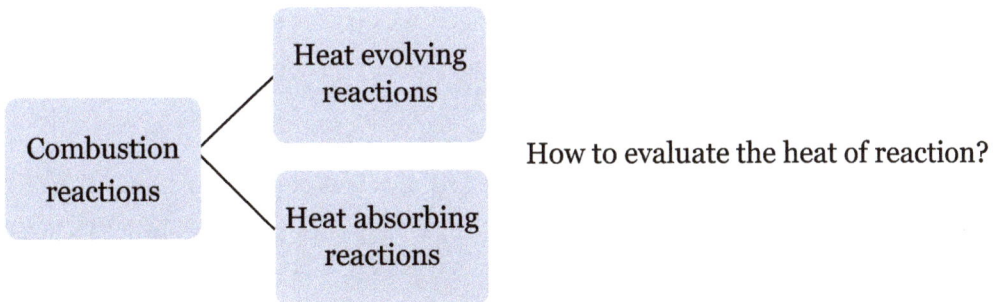

Consider the Burner as Shown Below

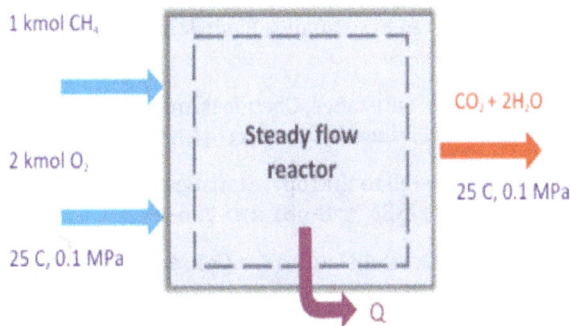

Assumption: (i) Negligible change in K.E. & P.E., (ii) No shaft work

$$Q = dH = H_p - H_R = \sum_{iP} n_{iP} h_{iP} - \sum_{iR} n_{iR} h_{iR} = \Delta H^{\circ}_{R,298}$$

Where,

H_R – Total enthalpy of reactants; HP – Total enthalpy of products

n_{iR} – No of moles of ith reactant; niP – No of moles of ith product

h_{iR} – Enthalpy of formation per unit mole of ith reactant

h_{iP} – Enthalpy of formation per unit mole of ith product

$\Delta H^{\circ}_{R,298}$ – Standard heat of reaction

H_p – Total enthalpy of products

n_{iP} – No of moles of ith product

Heat of Reaction Depends on Temperature

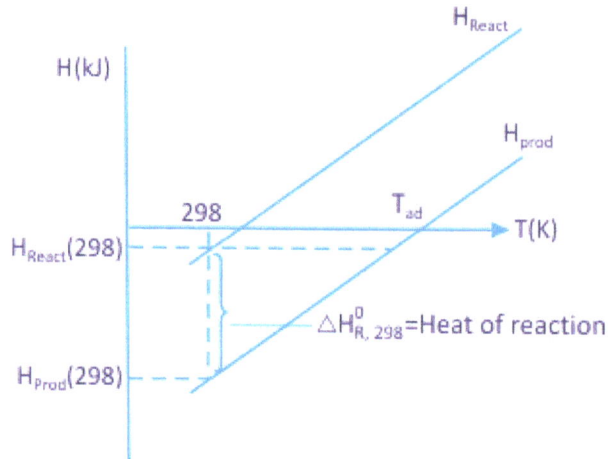

References

- Mahavira is dated 598 BC - 526 BC. Dundas, Paul; John Hinnels ed. (2002). The Jains. London: Routledge. ISBN 0-415-26606-8

- What's in a Name? Amount of Substance, Chemical Amount, and Stoichiometric Amount Carmen J. Giunta Journal of Chemical Education 2016 93 (4), 583-586 doi:10.1021/acs.jchemed.5b00690

- Ben-Naim, A. (2008). A Farewell to Entropy: Statistical Thermodynamics Based on Information, World Scientific, New Jersey, ISBN 978-981-270-706-2

- Robert D. Whitaker, "An Historical Note on the Conservation of Mass", Journal of Chemical Education, 52, 10, 658-659, Oct 75

- Chris Vuille; Serway, Raymond A.; Faughn, Jerry S. (2009). College physics. Belmont, CA: Brooks/Cole, Cengage Learning. p. 355. ISBN 0-495-38693-6

- "Energy Rules! Energy Conversion and the Laws of Thermodynamics - More About the First and Second Laws". Uwsp.edu. Archived from the original on 5 June 2010. Retrieved 12 September 2010

- Smith, J.M.; Van Ness, H.C.; Abbott, M.M. (2005). Introduction to Chemical Engineering Thermodynamics. McGraw Hill. ISBN 0-07-310445-0. OCLC 56491111

- "Conservation of Mass in Chemical Changes"Journal - Chemical Society, London, Vol.64, Part 2 Chemical Society (Great Britain)

- Iribarne, J.V., Godson, W.L. (1973/1981), Atmospheric Thermodynamics, second edition, D. Reidel, Kluwer Academic Publishers, Dordrecht, ISBN 90-277-1296-4, pages 48-49

3

Transport Phenomena in Combustion

The process of combustion can be classified into thermodynamics, fluid mechanics, heat and mass transfer, and chemical kinetics. The main governing laws of combution are Fick's law of diffusion, Newton's law of viscosity and Fourier's law of conduction. The aspects elucidated in this chapter are of vital importance, and provide a better understanding of transport phenomena in combustion.

Transport Phenomena

In engineering, physics and chemistry, the study of transport phenomena concerns the exchange of mass, energy, and momentum between observed and studied systems. While it draws from fields as diverse as continuum mechanics and thermodynamics, it places a heavy emphasis on the commonalities between the topics covered. Mass, momentum, and heat transport all share a very similar mathematical framework, and the parallels between them are exploited in the study of transport phenomena to draw deep mathematical connections that often provide very useful tools in the analysis of one field that are directly derived from the others.

While it draws its theoretical foundation from principles in a number of fields, most of the fundamental transport theory is a restatement of basic conservation laws.

The fundamental analyses in all three subfields of mass, heat, and momentum transfer are often grounded in the simple principle that the sum total of the quantities being studied must be conserved by the system and its environment. Thus, the different phenomena that lead to transport are each considered individually with the knowledge that the sum of their contributions must equal zero. This principle is useful for calculating many relevant quantities. For example, in fluid mechanics, a common use of transport analysis is to determine the velocity profile of a fluid flowing through a rigid volume.

Transport phenomena are ubiquitous throughout the engineering disciplines. Some of the most common examples of transport analysis in engineering are seen in the fields of process, chemical, biological, and mechanical engineering, but the subject is a fundamental component of the curriculum in all disciplines involved in any way with fluid mechanics, heat transfer, and mass transfer. It is now considered to be a part of the engineering discipline as much as thermodynamics, mechanics, and electromagnetism.

Transport phenomena encompass all agents of physical change in the universe. Moreover, they are considered to be fundamental building blocks which developed the universe, and which is responsible for the success of all life on earth. However, the scope here is limited to the relationship of transport phenomena to artificial engineered systems.

Overview

In physics, transport phenomena are all irreversible processes of statistical nature stemming from the random continuous motion of molecules, mostly observed in fluids. Every aspect of transport phenomena is grounded in two primary concepts : the conservation laws, and the constitutive equations. The conservation laws, which in the context of transport phenomena are formulated as continuity equations, describe how the quantity being studied must be conserved. The constitutive equations describe how the quantity in question responds to various stimuli via transport. Prominent examples include Fourier's Law of Heat Conduction and the Navier-Stokes equations, which describe, respectively, the response of heat flux to temperature gradients and the relationship between fluid flux and the forces applied to the fluid. These equations also demonstrate the deep connection between transport phenomena and thermodynamics, a connection that explains why transport phenomena are irreversible. Almost all of these physical phenomena ultimately involve systems seeking their lowest energy state in keeping with the principle of minimum energy. As they approach this state, they tend to achieve true thermodynamic equilibrium, at which point there are no longer any driving forces in the system and transport ceases. The various aspects of such equilibrium are directly connected to a specific transport: heat transfer is the system's attempt to achieve thermal equilibrium with its environment, just as mass and momentum transport move the system towards chemical and mechanical equilibrium.

Examples of transport processes include heat conduction (energy transfer), fluid flow (momentum transfer), molecular diffusion (mass transfer), radiation and electric charge transfer in semiconductors.

Transport phenomena have wide application. For example, in solid state physics, the motion and interaction of electrons, holes and phonons are studied under "transport phenomena". Another example is in biomedical engineering, where some transport phenomena of interest are thermoregulation, perfusion, and microfluidics. In chemical engineering, transport phenomena are studied in reactor design, analysis of molecular or diffusive transport mechanisms, and metallurgy.

The transport of mass, energy, and momentum can be affected by the presence of external sources:

- An odor dissipates more slowly (and may intensify) when the source of the odor remains present.

- The rate of cooling of a solid that is conducting heat depends on whether a heat source is applied.

- The gravitational force acting on a rain drop counteracts the resistance or drag imparted by the surrounding air.

Commonalities Among Phenomena

An important principle in the study of transport phenomena is analogy between phenomena.

Diffusion

There are some notable similarities in equations for momentum, energy, and mass transfer which can all be transported by diffusion, as illustrated by the following examples:

- Mass: the spreading and dissipation of odors in air is an example of mass diffusion.

- Energy: the conduction of heat in a solid material is an example of heat diffusion.

- Momentum: the drag experienced by a rain drop as it falls in the atmosphere is an example of momentum diffusion (the rain drop loses momentum to the surrounding air through viscous stresses and decelerates).

The molecular transfer equations of Newton's law for fluid momentum, Fourier's law for heat, and Fick's law for mass are very similar. One can convert from one transfer coefficient to another in order to compare all three different transport phenomena.

Comparison of Diffusion Phenomena		
Transported quantity	**Physical phenomenon**	**Equation**
Momentum	Viscosity (Newtonian fluid)	$\tau = -v \dfrac{\partial \rho v}{\partial x}$
Energy	Heat conduction (Fourier's law)	$\dfrac{q}{A} = -k \dfrac{dT}{dx}$
Mass	Molecular diffusion (Fick's law)	$J = -D \dfrac{\partial C}{\partial x}$

(Definitions of these formulas are given below).

A great deal of effort has been devoted in the literature to developing analogies among these three transport processes for turbulent transfer so as to allow prediction of one from any of the others. The Reynolds analogy assumes that the turbulent diffusivities are all equal and that the molecular diffusivities of momentum (μ/ρ) and mass (D_{AB})

are negligible compared to the turbulent diffusivities. When liquids are present and/ or drag is present, the analogy is not valid. Other analogies, such as von Karman's and Prandtl's, usually result in poor relations.

The most successful and most widely used analogy is the Chilton and Colburn J-factor analogy. This analogy is based on experimental data for gases and liquids in both the laminar and turbulent regimes. Although it is based on experimental data, it can be shown to satisfy the exact solution derived from laminar flow over a flat plate. All of this information is used to predict transfer of mass.

Onsager Reciprocal Relations

In fluid systems described in terms of temperature, matter density, and pressure, it is known that temperature differences lead to heat flows from the warmer to the colder parts of the system; similarly, pressure differences will lead to matter flow from high-pressure to low-pressure regions (a "reciprocal relation"). What is remarkable is the observation that, when both pressure and temperature vary, temperature differences at constant pressure can cause matter flow (as in convection) and pressure differences at constant temperature can cause heat flow. Perhaps surprisingly, the heat flow per unit of pressure difference and the density (matter) flow per unit of temperature difference are equal.

This equality was shown to be necessary by Lars Onsager using statistical mechanics as a consequence of the time reversibility of microscopic dynamics. The theory developed by Onsager is much more general than this example and capable of treating more than two thermodynamic forces at once.

Momentum Transfer

In momentum transfer, the fluid is treated as a continuous distribution of matter. The study of momentum transfer, or fluid mechanics can be divided into two branches: fluid statics (fluids at rest), and fluid dynamics (fluids in motion). When a fluid is flowing in the x direction parallel to a solid surface, the fluid has x-directed momentum, and its concentration is $v_x \rho$. By random diffusion of molecules there is an exchange of molecules in the z direction. Hence the x-directed momentum has been transferred in the z-direction from the faster- to the slower-moving layer. The equation for momentum transport is Newton's Law of Viscosity written as follows:

$$\tau_{zx} = -v \frac{\partial \rho v_x}{\partial z}$$

where τ_{zx} is the flux of x-directed momentum in the z direction, v is μ/ρ, the momentum diffusivity, z is the distance of transport or diffusion, ρ is the density, and μ is the viscosity. Newtons Law is the simplest relationship between the flux of momentum and the velocity gradient.

Mass Transfer

When a system contains two or more components whose concentration vary from point to point, there is a natural tendency for mass to be transferred, minimizing any concentration difference within the system. Mass Transfer in a system is governed by Fick's First Law: 'Diffusion flux from higher concentration to lower concentration is proportional to the gradient of the concentration of the substance and the diffusivity of the substance in the medium.' Mass transfer can take place due to different driving forces. Some of them are:

- Mass can be transferred by the action of a pressure gradient(pressure diffusion)

- Forced diffusion occurs because of the action of some external force

- Diffusion can be caused by temperature gradients (thermal diffusion)

- Diffusion can be caused by differences in chemical potential

This can be compared to Fick's Law of Diffusion:

$$J_{Ay} = -D_{AB} \frac{\partial Ca}{\partial y}$$

where D is the diffusivity constant.

Energy Transfer

All processes in engineering involve the transfer of energy. Some examples are the heating and cooling of process streams, phase changes, distillations, etc. The basic principle is the first law of thermodynamics which is expressed as follows for a static system:

$$q = -k \frac{dT}{dx}$$

The net flux of energy through a system equals the conductivity times the rate of change of temperature with respect to position.

For other systems that involve either turbulent flow, complex geometries or difficult boundary conditions another equation would be easier to use:

$$Q = h \cdot A \cdot \Delta T$$

where A is the surface area, : ΔT is the temperature driving force, Q is the heat flow per unit time, and h is the heat transfer coefficient.

Within heat transfer, two types of convection can occur:

Forced convection can occur in both laminar and turbulent flow. In the situation of laminar flow in circular tubes, several dimensionless numbers are used such as Nusselt number, Reynolds number, and Prandtl. The commonly used equation is:

$$Nu_a = \frac{h_a D}{k}$$

Natural or free convection is a function of Grashof and Prandtl numbers. The complexities of free convection heat transfer make it necessary to mainly use empirical relations from experimental data.

Heat transfer is analyzed in packed beds, reactors and heat exchangers.

Laws of Transport Phenomenon

Newton's Law of Viscosity

- Two parallel plates-separated by 'Y'

- Lower plate is fixed

- At t< 0; system is at rest

- At t=0; upper plate is moved

- Velocity of plate: V_x

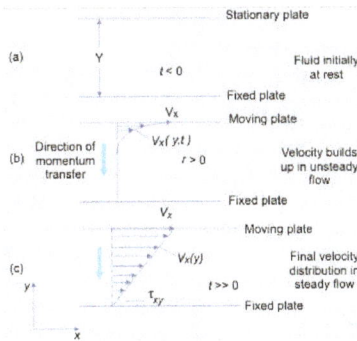

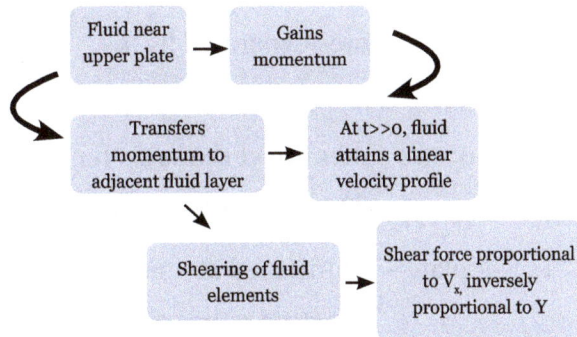

$$\tau_{yx} = \frac{F}{A} \propto \frac{V_x}{Y}$$

Newton's Law of Viscosity can be Expressed as

$$\tau_{yx} = -\mu \frac{dV_x}{d_y}$$

Where,

μ is dynamic viscosity (kg/ms)

dVx/dy is the shear strain rate

-ve sign: momentum flux in the direction of decreasing velocity

Viscosity for gases:

  - Independent of pressure
  - Temperature dependent

$$\mu \propto T^{0.7}$$

Viscosity for liquids:

  - Decrease with increase in temperature

Newtonian Fluids:

Gases and liquids which follow Newton's law of viscosity

Momentum transfer takes place from the region of higher velocity to lower velocity

Non- Newtonian Fluids:

Fluids which do not follow Newton's law of viscosity

Fourier's Law of Heat Conduction

  - Two parallel plates-separated by 'Y'

  - Lower plate is fixed

  - At t < 0; two plates are at the same temperature

  - At t = 0; upper plate is suddenly heated ($T_1 > T_0$)

  - Lower plate – Maintained at temperature T_0

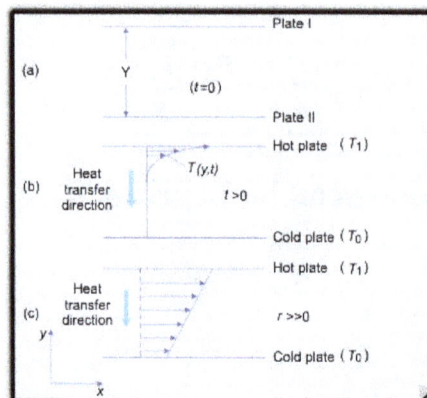

$$\dot{q}'' = \frac{Q}{A} \propto \frac{T_1 - T_0}{Y}$$

$$q'' = -k\frac{dT}{d_y}$$

Upper plate maintained at high temperature → Heat transferred to fluid at the vicinity of the plate

Heat transferred to adjacent molecules ← Vibration of molecules

Over a period of time, temperature attains a steady state → Heat-flux proportional to temperature difference, inversely proportional to Y

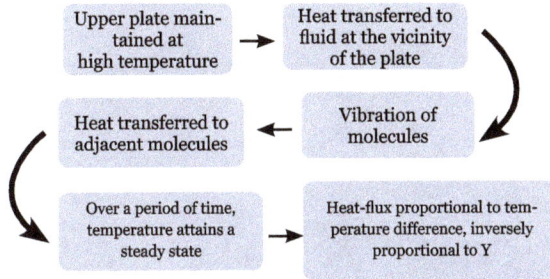

Fick's Law of Species Diffusion

- Two parallel plates-separated by 'Y'

- Upper plate is maintained wet

- Lower plate is kept dry (Dehydrating agent)

- Water vapour evaporates at upper plate

- Partial pressure of water vapour is maintained at saturated vapour pressure of water

- Thus concentration gradient exists between the two plates

- Mass flux-proportional to concentration, CA inversely proportional to distance Y

$$m''_w \propto \frac{C_A}{Y_D}$$

Differential form of Fick's law

$$m''_A = -\rho D_{AB}\frac{dY_A}{d_Y}$$

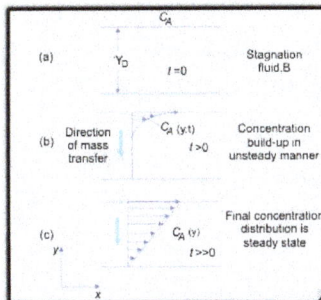

(a) Y_D C_A $t = 0$ Stagnation fluid.B

(b) Direction of mass transfer $C_A (y,t)$ $t > 0$ Concentration build-up in unsteady manner

(c) y_A $C_A (y)$ $t >> 0$ Final concentration distribution is steady state

D_{AB} Binary diffusitivity of species A through B

Fick's Laws of Diffusion

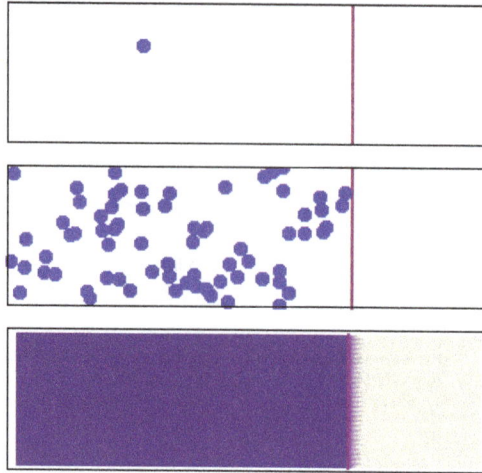

Molecular diffusion from a microscopic and macroscopic point of view. Initially, there are solute molecules on the left side of a barrier (purple line) and none on the right. The barrier is removed, and the solute diffuses to fill the whole container. Top: A single molecule moves around randomly. Middle: With more molecules, there is a clear trend where the solute fills the container more and more uniformly. Bottom: With an enormous number of solute molecules, randomness becomes undetectable: The solute appears to move smoothly and systematically from high-concentration areas to low-concentration areas. This smooth flow is described by Fick's laws.

Fick's laws of diffusion describe diffusion and were derived by Adolf Fick in 1855. They can be used to solve for the diffusion coefficient, D. Fick's first law can be used to derive his second law which in turn is identical to the diffusion equation.

Fick's First Law

Fick's first law relates the diffusive flux to the concentration under the assumption of steady state. It postulates that the flux goes from regions of high concentration to regions of low concentration, with a magnitude that is proportional to the concentration gradient (spatial derivative), or in simplistic terms the concept that a solute will move from a region of high concentration to a region of low concentration across a concentration gradient. In one (spatial) dimension, the law is:

$$J = -D\frac{d\varphi}{dx}$$

where

- J is the "diffusion flux," of which the dimension is amount of substance per unit area per unit time, so it is expressed in such units as mol m^{-2} s^{-1}. J measures the amount of substance that will flow through a unit area during a unit time interval.

- D is the diffusion coefficient or diffusivity. Its dimension is area per unit time, so typical units for expressing it would be m²/s.

- φ (for ideal mixtures) is the concentration, of which the dimension is amount of substance per unit volume. It might be expressed in units of mol/m³.

- x is position, the dimension of which is length. It might thus be expressed in the unit m.

D is proportional to the squared velocity of the diffusing particles, which depends on the temperature, viscosity of the fluid and the size of the particles according to the Stokes-Einstein relation. In dilute aqueous solutions the diffusion coefficients of most ions are similar and have values that at room temperature are in the range of 0.6×10^{-9} to 2×10^{-9} m²/s. For biological molecules the diffusion coefficients normally range from 10^{-11} to 10^{-10} m²/s.

In two or more dimensions we must use ∇, the del or gradient operator, which generalises the first derivative, obtaining

$$\mathbf{J} = -D\nabla\varphi$$

where J denotes the diffusion flux vector.

The driving force for the one-dimensional diffusion is the quantity $-\dfrac{\partial\varphi}{\partial X}$, which for ideal mixtures is the concentration gradient. In chemical systems other than ideal solutions or mixtures, the driving force for diffusion of each species is the gradient of chemical potential of this species. Then Fick's first law (one-dimensional case) can be written as:

$$J_i = -\frac{Dc_i}{RT}\frac{\partial\mu_i}{\partial x}$$

where the index i denotes the ith species, c is the concentration (mol/m³), R is the universal gas constant (J/K/mol), T is the absolute temperature (K), and μ is the chemical potential (J/mol).

If the primary variable is mass fraction (y_i, given, for example, in kg/kg), then the equation changes to:

$$J_i = -\rho D\nabla y_i$$

where ρ is the fluid density (for example, in kg/m³). Note that the density is outside the gradient operator.

Fick's Second Law

Fick's second law predicts how diffusion causes the concentration to change with time. It is a partial differential equation which in one dimension reads:

$$\frac{\partial \varphi}{\partial t} = D\frac{\partial^2 \varphi}{\partial x^2}$$

where

- φ is the concentration in dimensions of [(amount of substance) length^{-3}], example mol/m³; $\varphi = \varphi(x,t)$ is a function that depends on location x and time t

- t is time [s]

- D is the diffusion coefficient in dimensions of [length^2 time^{-1}], example m²/s

- x is the position [length], example m

In two or more dimensions we must use the Laplacian $\mathbf{\Delta} = \nabla^2$, which generalises the second derivative, obtaining the equation

$$\frac{\partial \varphi}{\partial t} = D\Delta\varphi$$

Derivation

Fick's second law is a special case of the convection–diffusion equation in which there is no advective flux and no net volumetric source. It can be derived from the continuity equation:

$$\frac{\partial \varphi}{\partial t} + \nabla \cdot \vec{j} = R,$$

where $\vec{j}$ is the total flux and R is a net volumetric source for φ. The only source of flux in this situation is assumed to be diffusive flux:

$$\vec{j}_{\text{diffusion}} = -D\nabla\varphi$$

Plugging the definition of diffusive flux to the continuity equation and assuming there is no source (R = 0), we arrive at Fick's second law:

$$\frac{\partial \varphi}{\partial t} = D\frac{\partial^2 \varphi}{\partial x^2}$$

If flux were the result of both diffusive flux and advective flux, the convection–diffusion equation is the result.

Example Solution in One Dimension: Diffusion Length

A simple case of diffusion with time t in one dimension (taken as the x-axis) from a boundary located at position $x = 0$, where the concentration is maintained at a value n_0 is

$$n(x,t) = n_0 \mathrm{erfc}\left(\frac{x}{2\sqrt{Dt}}\right).$$

where erfc is the complementary error function. This is the case when corrosive gases diffuse through the oxidative layer towards the metal surface (if we assume that concentration of gases in the environment is constant and the diffusion space (i. e., corrosion product layer) is *semi-infinite* – starting at 0 at the surface and spreading infinitely deep in the material). If, in its turn, the diffusion space is *infinite* (lasting both through the layer with $n(x,0) = 0$, $x > 0$ and that with $n(x,0) = n_0$, $x \leq 0$), then the solution is amended only with coefficient $\frac{1}{2}$ in front of n_0 (this might seem obvious, as the diffusion now occurs in both directions). This case is valid when some solution with concentration n_0 is put in contact with a layer of pure solvent. (Bokstein, 2005) The length $2\sqrt{Dt}$ is called the *diffusion length* and provides a measure of how far the concentration has propagated in the x-direction by diffusion in time t (Bird, 1976).

As a quick approximation of the error function, the first 2 terms of the Taylor series can be used:

$$n(x,t) = n_0 \left[1 - 2\left(\frac{x}{2\sqrt{Dt\pi}}\right)\right]$$

If D is time-dependent, the diffusion length becomes

$$2\sqrt{\int_0^t D(t')dt'}\,.$$

This idea is useful for estimating a diffusion length over a heating and cooling cycle, where D varies with temperature.

Generalizations

1. In *inhomogeneous media*, the diffusion coefficient varies in space, $D = D(x)$. This dependence does not affect Fick's first law but the second law changes:

$$\frac{\partial \varphi(x,t)}{\partial t} = \nabla \cdot (D(x)\nabla \varphi(x,t)) = D(x)\Delta \varphi(x,t) + \sum_{i=1}^{3} \frac{\partial D(x)}{\partial x_i} \frac{\partial \varphi(x,t)}{\partial x_i}$$

2. In *anisotropic media*, the diffusion coefficient depends on the direction. It is a symmetric tensor $D = D_{ij}$. Fick's first law changes to

$$J = -D\nabla\varphi,$$

it is the product of a tensor and a vector:

$$J_i = -\sum_{j=1}^{3} D_{ij} \frac{\partial\varphi}{\partial x_j}.$$

For the diffusion equation this formula gives

$$\frac{\partial\varphi(x,t)}{\partial t} = \nabla\cdot(D\nabla\varphi(x,t)) = \sum_{i=1}^{3}\sum_{j=1}^{3} D_{ij} \frac{\partial^2\varphi(x,t)}{\partial x_i\partial x_j}.$$

The symmetric matrix of diffusion coefficients D_{ij} should be positive definite. It is needed to make the right hand side operator elliptic.

3. For *inhomogeneous anisotropic media* these two forms of the diffusion equation should be combined in

$$\frac{\partial\varphi(x,t)}{\partial t} = \nabla\cdot(D(x)\nabla\varphi(x,t)) = \sum_{i,j=1}^{3}\left(D_{ij}(x)\frac{\partial^2\varphi(x,t)}{\partial x_i\partial x_j} + \frac{\partial D_{ij}(x)}{\partial x_i}\frac{\partial\varphi(x,t)}{\partial x_j} \right).$$

4. The approach based on Einstein's mobility and Teorell formula gives the following generalization of Fick's equation for the *multicomponent diffusion* of the perfect components:

$$\frac{\partial\varphi_i}{\partial t} = \sum_j \nabla\cdot\left(D_{ij} \frac{\varphi_i}{\varphi_j}\nabla\varphi_j \right).$$

where φ_i are concentrations of the components and D_{ij} is the matrix of coefficients. Here, indices i, j are related to the various components and not to the space coordinates.

The Chapman–Enskog formulae for diffusion in gases include exactly the same terms. It should be stressed that these physical models of diffusion are different from the test models $\partial_t\varphi_i = \Sigma_j D_{ij}\Delta\varphi_j$ which are valid for very small deviations from the uniform equilibrium. Earlier, such terms were introduced in the Maxwell–Stefan diffusion equation.

For anisotropic multicomponent diffusion coefficients one needs a rank-four tensor, for example $D_{ij,\alpha\beta}$, where i, j refer to the components and $\alpha, \beta = 1, 2, 3$ correspond to the space coordinates.

Applications

Equations based on Fick's law have been commonly used to model transport processes in foods, neurons, biopolymers, pharmaceuticals, porous soils, population dynamics, nuclear materials, semiconductor doping process, etc. Theory of all voltammetric methods is based on solutions of Fick's equation. A large amount of experimental research in polymer science and food science has shown that a more general approach is required to describe transport of components in materials undergoing glass transition. In the vicinity of glass transition the flow behavior becomes "non-Fickian". It can be shown that the Fick's law can be obtained from the Maxwell-Stefan equations of multi-component mass transfer. The Fick's law is limiting case of the Maxwell-Stefan equations, when the mixture is extremely dilute and every chemical species is interacting only with the bulk mixture and not with other species. To account for the presence of multiple species in a non-dilute mixture, several variations of the Maxwell-Stefan equations are used.

Biological Perspective

The first law gives rise to the following formula:

$$\text{Flux} = -P(c_2 - c_1)$$

in which,

- P is the permeability, an experimentally determined membrane "conductance" for a given gas at a given temperature.

- $c_2 - c_1$ is the difference in concentration of the gas across the membrane for the direction of flow (from c_1 to c_2).

Fick's first law is also important in radiation transfer equations. However, in this context it becomes inaccurate when the diffusion constant is low and the radiation becomes limited by the speed of light rather than by the resistance of the material the radiation is flowing through. In this situation, one can use a flux limiter.

The exchange rate of a gas across a fluid membrane can be determined by using this law together with Graham's law.

Fick's Flow in Liquids

When two miscible liquids are brought into contact, and diffusion takes place, the macroscopic (or average) concentration evolves following Fick's law. On a mesoscopic scale, that is, between the macroscopic scale described by Fick's law and molecular scale, where molecular random walks take place, fluctuations cannot be neglected. Such situations can be successfully modeled with Landau-Lifshitz fluctuating hydrodynamics.

In this theoretical framework, diffusion is due to fluctuations whose dimensions range from the molecular scale to the macroscopic scale.

In particular, fluctuating hydrodynamic equations include a Fick's flow term, with a given diffusion coefficient, along with hydrodynamics equations and stochastic terms describing fluctuations. When calculating the fluctuations with a perturbative approach, the zero order approximation is Fick's law. The first order gives the fluctuations, and it comes out that fluctuations contribute to diffusion. This represents somehow a tautology, since the phenomena described by a lower order approximation is the result of a higher approximation: this problem is solved only by renormalizing the fluctuating hydrodynamics equations.

Semiconductor Fabrication Applications

Integrated circuit fabrication technologies, model processes like CVD, thermal oxidation, wet oxidation, doping, etc. use diffusion equations obtained from Fick's law.

In certain cases, the solutions are obtained for boundary conditions such as constant source concentration diffusion, limited source concentration, or moving boundary diffusion (where junction depth keeps moving into the substrate).

Derivation of Fick's Laws

Fick's First Law

In one dimension, the following derivation is based on a similar argument made in Berg 1977.

Consider a collection of particles performing a random walk in one dimension with length scale Δx and time scale Δt. Let $N(x,t)$ be the number of particles at position x at time t.

At a given time step, half of the particles would move left and half would move right. Since half of the particles at point x move right and half of the particles at point $x + \Delta x$ move left, the net movement to the right is:

$$-\tfrac{1}{2}\left[N(x+\Delta x,t) - N(x,t)\right]$$

The flux, J, is this net movement of particles across some area element of area a, normal to the random walk during a time interval Δt. Hence we may write:

$$J = -\frac{1}{2}\left[\frac{N(x+\Delta x,t)}{a\Delta t} - \frac{N(x,t)}{a\Delta t}\right]$$

Multiplying the top and bottom of the righthand side by $(\Delta x)^2$ and rewriting, we obtain:

$$J = -\frac{(\Delta x)^2}{2\Delta t}\left[\frac{N(x+\Delta x,t)}{a(\Delta x)^2} - \frac{N(x,t)}{a(\Delta x)^2}\right]$$

We note that concentration is defined as particles per unit volume, and hence

$$\varphi(x,t) = \frac{N(x,t)}{a\Delta x}.$$

In addition, $\frac{(\Delta X)^2}{2\Delta t}$ is the definition of the diffusion constant in one dimension, D.
Thus our expression simplifies to:

$$J = -D\left[\frac{\varphi(x+\Delta x,t)}{\Delta x} - \frac{\varphi(x,t)}{\Delta x}\right]$$

In the limit where Δx is infinitesimal, the righthand side becomes a space derivative:

$$J = -D\frac{\partial\varphi}{\partial x}$$

Fick's Second Law

Fick's second law can be derived from Fick's First law and the mass conservation in absence of any chemical reactions:

$$\frac{\partial\varphi}{\partial t} + \frac{\partial}{\partial x}J = 0 \Rightarrow \frac{\partial\varphi}{\partial t} - \frac{\partial}{\partial x}\left(D\frac{\partial}{\partial x}\varphi\right) = 0$$

Assuming the diffusion coefficient D to be a constant, we can exchange the orders of the differentiation and multiply by the constant:

$$\frac{\partial}{\partial x}\left(D\frac{\partial}{\partial x}\varphi\right) = D\frac{\partial}{\partial x}\frac{\partial}{\partial x}\varphi = D\frac{\partial^2\varphi}{\partial x^2}$$

and, thus, receive the form of the Fick's equations as was stated above.

For the case of diffusion in two or more dimensions Fick's Second Law becomes

$$\frac{\partial\varphi}{\partial t} = D\nabla^2\varphi$$
,

which is analogous to the heat equation.

If the diffusion coefficient is not a constant, but depends upon the coordinate and/or concentration, Fick's Second Law yields

$$\frac{\partial \varphi}{\partial t} = \nabla \cdot (D \nabla \varphi)$$

An important example is the case where φ is at a steady state, i.e. the concentration does not change by time, so that the left part of the above equation is identically zero. In one dimension with constant D, the solution for the concentration will be a linear change of concentrations along x. In two or more dimensions we obtain

$$\nabla^2 \varphi = 0$$

which is Laplace's equation, the solutions to which are referred to by mathematicians as harmonic functions.

History

In 1855, physiologist Adolf Fick first reported his now-well-known laws governing the transport of mass through diffusive means. Fick's work was inspired by the earlier experiments of Thomas Graham, which fell short of proposing the fundamental laws for which Fick would become famous. The Fick's law is analogous to the relationships discovered at the same epoch by other eminent scientists: Darcy's law (hydraulic flow), Ohm's law (charge transport), and Fourier's Law (heat transport).

Fick's experiments (modeled on Graham's) dealt with measuring the concentrations and fluxes of salt, diffusing between two reservoirs through tubes of water. It is notable that Fick's work primarily concerned diffusion in fluids, because at the time, diffusion in solids was not considered generally possible. Today, Fick's Laws form the core of our understanding of diffusion in solids, liquids, and gases (in the absence of bulk fluid motion in the latter two cases). When a diffusion process does *not* follow Fick's laws (which does happen), it is referred to as *non-Fickian*, in that they are exceptions that "prove" the importance of the general rules that Fick outlined in 1855.

Transport Properties for Gas Mixture

Viscosity of Gas Mixture

Wassilijewa equation

$$\mu_{mix} = \sum_{i=1}^{N} \frac{X_i \mu_i}{\sum_{j=1}^{N} X_i A_{ij}}$$

Mason and Saxena modification

$$A_{ij} = \frac{\left[1 + \left(\frac{\mu_i}{\mu_j} \right)^{0.5} \left(\frac{MW_i}{MW_j} \right)^{0.25} \right]}{\left[8 \left(1 + \left(\frac{MW_i}{MW_j} \right) \right)^{0.5} \right]}$$

μ_i is the viscosity of the pure component

X_i is the mole fraction of the ith component

Thermal Conductivity of Gas Mixture

Wassilijewa equation   Mason and Saxena modification

$$k_{mix} = \sum_{j=1}^{N} \frac{X_i k_i}{\sum_{j=1}^{N} X_i A_{ij}}$$

$$A_{ij} = \frac{\left[1 + \left(\dfrac{k_i}{k_j} \right)^{0.5} \left(\dfrac{MW_i}{MW_j} \right)^{0.25} \right]^z}{\left[8 \left(1 + \left(\dfrac{MW_i}{MW_j} \right) \right)^{0.5} \right]}$$

k_i is the thermal conductivity of the pure component

Diffusion Coefficient of Any Component in a Gas Mixture

- Wilke equation

$$D_i = \frac{1 - X_i}{\sum_{j=1 \neq i}^{N} \dfrac{X_j}{D_{ij}}}$$

$$D_i = 262.8 \times 10^{-9} \frac{\sqrt{T^3 \dfrac{MW_i + MW_j}{2 MW_i MW_j}}}{P \sigma_{ij}^2 \Omega_D}$$

Where,

σ_{ij} is the collision diameter in A^0

P is the pressure (Bar)

MW_i is the molecular weight of the components

Ω_D is the collision integral $T_{ij}^* = \sqrt{T_i^* T_j^*}$

$$\Omega_D = (44.54 \times T_{ij}^{*-4.909} + 1.911 \times T_{ij}^{*1.575})^{0.1} \qquad T_i^* = \frac{T}{\varepsilon_0 / k_B}$$

k_B = Boltzman's constant

ε_0 = Intermolecular potential

Mass Conservation Equation

Principle of Mass Conservation

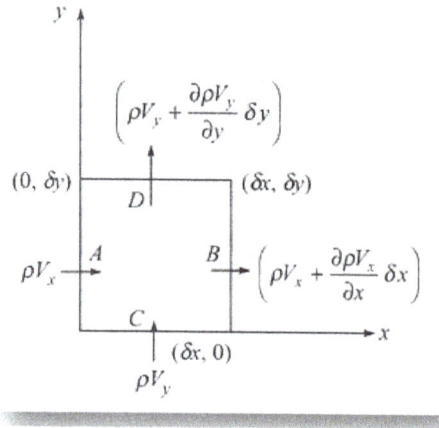

$$\begin{Bmatrix} \text{Rate of mass accumulation} \\ \text{in fluid element} \end{Bmatrix} =$$

$$\begin{Bmatrix} \text{Rate of mass into} \\ \text{fluid element} \end{Bmatrix} - \begin{Bmatrix} \text{Rate of mass out of} \\ \text{fluid element} \end{Bmatrix} \qquad (1)$$

Rate of accumulation in fluid element $= \dfrac{\partial \rho}{\partial t}(\delta x \times \delta y \times 1)$ \qquad (2)

Rate of mass in fluid element across face A $= \rho V_x(\delta y \times 1)$ \qquad (3)

Rate of mass leaving fluid element across face B $= \left(\rho V_x + \dfrac{\partial(\rho V_x)}{\partial x}\delta x \right) \times (\delta y \times 1)$ \qquad (4)

The net efflux in x-direction $= \dfrac{\partial(\rho V_x)}{\partial x} \times (\delta x \times \delta y)$ \qquad (5)

The net efflux in y-direction $= \dfrac{\partial(\rho V_y)}{\partial y} \times (\delta x \times \delta y)$ \qquad (6)

Substituting (2), (5) and (6) in (1)

$$\frac{\partial \rho}{\partial t} + \frac{\partial(\rho V_x)}{\partial x} + \frac{\partial(\rho V_y)}{\partial y} = 0$$

Differential form of continuity equation

In vector notation,

$$\frac{\partial \rho}{\partial t} + \nabla . \rho V = 0$$

Where, ∇ is the gradient operator

$\nabla . (\rho V)$ is the divergence of ρV

For Incompressible Flow

$$\frac{\partial V_x}{\partial x} + \frac{\partial V_y}{\partial y} = 0$$

Momentum

In classical mechanics, linear momentum, translational momentum, or simply momentum (pl. momenta; SI unit kg · m/s) is the product of the mass and velocity of an object, quantified in kilogram-meters per second. It is dimensionally equivalent to impulse, the product of force and time, quantified in newton-seconds. Newton's second law of motion states that the change in linear momentum of a body is equal to the net impulse acting on it. For example, a heavy truck moving rapidly has a large momentum, and it takes a large or prolonged force to get the truck up to this speed, and would take a similarly large or prolonged force to bring it to a stop. If the truck were lighter, or moving more slowly, then it would have less momentum and therefore require less impulse to start or stop.

Like velocity, linear momentum is a vector quantity, possessing a direction as well as a magnitude:

$$\mathbf{p} = m\mathbf{v},$$

where p is the three-dimensional vector stating the object's momentum in the three directions of three-dimensional space, v is the three-dimensional velocity vector giving the object's rate of movement in each direction, and m is the object's mass.

Linear momentum is also a *conserved* quantity, meaning that if a closed system is not affected by external forces, its total linear momentum cannot change.

In classical mechanics, conservation of linear momentum is implied by Newton's laws. It also holds in special relativity (with a modified formula) and, with appropriate definitions, a (generalized) linear momentum conservation law holds in electrodynamics, quantum mechanics, quantum field theory, and general relativity. It is ultimately an expression of one of the fundamental symmetries of space and time, that of translational symmetry.

Linear momentum depends on frame of reference. Observers in different frames would find different values of linear momentum of a system. But each would observe that the value of linear momentum does not change with time, provided the system is isolated.

Newtonian Mechanics

Momentum has a direction as well as magnitude. Quantities that have both a magnitude and a direction are known as vector quantities. Because momentum has a direction, it can be used to predict the resulting direction of objects after they collide, as well as their speeds. Below, the basic properties of momentum are described in one dimension. The vector equations are almost identical to the scalar equations.

Single Particle

The momentum of a particle is traditionally represented by the letter p. It is the product of two quantities, the mass (represented by the letter m) and velocity (v):

$$p = mv.$$

The units of momentum are the product of the units of mass and velocity. In SI units, if the mass is in kilograms and the velocity in meters per second then the momentum is in kilogram meters/second (kg m/s). In cgs units, if the mass is in grams and the velocity in centimeters per second, then the momentum is in gram centimeters/second (g cm/s).

Being a vector, momentum has magnitude and direction. For example, a 1 kg model airplane, traveling due north at 1 m/s in straight and level flight, has a momentum of 1 kg m/s due north measured from the ground.

Many Particles

The momentum of a system of particles is the sum of their momenta. If two particles have masses m_1 and m_2, and velocities v_1 and v_2, the total momentum is

$$p = p_1 + p_2$$
$$= m_1 v_1 + m_2 v_2.$$

The momenta of more than two particles can be added more generally with the following:

$$p = \sum_i m_i v_i$$

A system of particles has a center of mass, a point determined by the weighted sum of their positions:

$$r_{cm} = \frac{m_1 r_1 + m_2 r_2 + \cdots}{m_1 + m_2 + \cdots} = \frac{\sum_i m_i r_i}{\sum_i m_i}.$$

If all the particles are moving, the center of mass will generally be moving as well (unless the system is in pure rotation around it). If the center of mass is moving at velocity v_{cm}, the momentum is:

$$p = m v_{cm}.$$

This is known as Euler's first law.

Relation to Force

If a force F is applied to a particle for a time interval Δt, the momentum of the particle changes by an amount

$$\Delta p = F \Delta t.$$

In differential form, this is Newton's second law; the rate of change of the momentum of a particle is proportional to the force F acting on it,

$$F = \frac{dp}{dt}.$$

If the force depends on time, the change in momentum (or impulse J) between times t_1 and t_2 is

$$\Delta p = J = \int_{t_1}^{t_2} F(t) dt.$$

Impulse is measured in the derived units of the newton second (1 N s = 1 kg m/s) or dyne second (1 dyne s = 1 g m/s)

Under the assumption of constant mass m, it is equivalent to write

$$F = m \frac{dv}{dt} = ma,$$

so the force is equal to mass times acceleration.

Example: A model airplane of 1 kg accelerates from rest to a velocity of 6 m/s due north in 2 s. The net force required to produce this acceleration is 3 newtons due north. The change in momentum is 6 kg m/s. The rate of change of momentum is 3 (kg m/s)/s = 3 N.

Conservation

A Newton's cradle demonstrates conservation of momentum.

In a closed system (one that does not exchange any matter with its surroundings and is not acted on by external forces) the total momentum is constant. This fact, known as the *law of conservation of momentum*, is implied by Newton's laws of motion. Suppose, for example, that two particles interact. Because of the third law, the forces between them are equal and opposite. If the particles are numbered 1 and 2, the second law states that $F_1 = \dfrac{dp_1}{dt}$ and $F_2 = \dfrac{dp_2}{dt}$. Therefore,

$$\frac{dp_1}{dt} = -\frac{dp_2}{dt},$$

with the negative sign indicating that the forces oppose. Equivalently,

$$\frac{d}{dt}(p_1 + p_2) = 0.$$

If the velocities of the particles are u_1 and u_2 before the interaction, and afterwards they are v_1 and v_2, then

$$m_1 u_1 + m_2 u_2 = m_1 v_1 + m_2 v_2.$$

This law holds no matter how complicated the force is between particles. Similarly, if there are several particles, the momentum exchanged between each pair of particles adds up to zero, so the total change in momentum is zero. This conservation law applies to all interactions, including collisions and separations caused by explosive forces. It can also be generalized to situations where Newton's laws do not hold, for example in the theory of relativity and in electrodynamics.

Dependence on Reference Frame

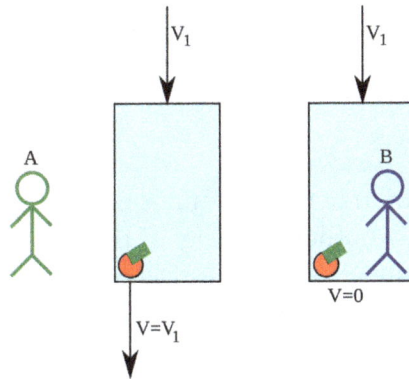

Newton's apple in Einstein's elevator. In person A's frame of reference,
the apple has non-zero velocity and momentum. In the elevator's and
person B's frames of reference, it has zero velocity and momentum.

Momentum is a measurable quantity, and the measurement depends on the motion of the observer. For example: if an apple is sitting in a glass elevator that is descending, an outside observer, looking into the elevator, sees the apple moving, so, to that observer, the apple has a non-zero momentum. To someone inside the elevator, the apple does not move, so, it has zero momentum. The two observers each have a frame of reference, in which, they observe motions, and, if the elevator is descending steadily, they will see behavior that is consistent with those same physical laws.

Suppose a particle has position x in a stationary frame of reference. From the point of view of another frame of reference, moving at a uniform speed u, the position (represented by a primed coordinate) changes with time as

$$x' = x - ut.$$

This is called a Galilean transformation. If the particle is moving at speed $\frac{dx}{dt} = v$ in the first frame of reference, in the second, it is moving at speed

$$v' = \frac{dx'}{dt} = v - u.$$

Since u does not change, the accelerations are the same:

$$a' = \frac{dv'}{dt} = a.$$

Thus, momentum is conserved in both reference frames. Moreover, as long as the force has the same form, in both frames, Newton's second law is unchanged. Forces such as Newtonian gravity, which depend only on the scalar distance between objects, satisfy this criterion. This independence of reference frame is called Newtonian relativity or Galilean invariance.

A change of reference frame, can, often, simplify calculations of motion. For example, in a collision of two particles, a reference frame can be chosen, where, one particle begins at rest. Another, commonly used reference frame, is the center of mass frame - one that is moving with the center of mass. In this frame, the total momentum is zero.

Application to Collisions

By itself, the law of conservation of momentum is not enough to determine the motion of particles after a collision. Another property of the motion, kinetic energy, must be known. This is not necessarily conserved. If it is conserved, the collision is called an *elastic collision*; if not, it is an *inelastic collision*.

Elastic Collisions

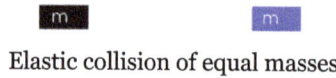

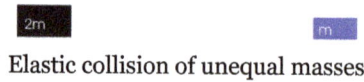

Elastic collision of equal masses

Elastic collision of unequal masses

An elastic collision is one in which no kinetic energy is lost. Perfectly elastic "collisions" can occur when the objects do not touch each other, as for example in atomic or nuclear scattering where electric repulsion keeps them apart. A slingshot maneuver of a satellite around a planet can also be viewed as a perfectly elastic collision from a distance. A collision between two pool balls is a good example of an *almost* totally elastic collision, due to their high rigidity; but when bodies come in contact there is always some dissipation.

A head-on elastic collision between two bodies can be represented by velocities in one dimension, along a line passing through the bodies. If the velocities are u_1 and u_2 before the collision and v_1 and v_2 after, the equations expressing conservation of momentum and kinetic energy are:

$$m_1 u_1 + m_2 u_2 = m_1 v_1 + m_2 v_2$$
$$\tfrac{1}{2} m_1 u_1^2 + \tfrac{1}{2} m_2 u_2^2 = \tfrac{1}{2} m_1 v_1^2 + \tfrac{1}{2} m_2 v_2^2.$$

A change of reference frame can often simplify the analysis of a collision. For example, suppose there are two bodies of equal mass m, one stationary and one approaching the other at a speed v (as in the figure). The center of mass is moving at speed $\dfrac{v}{2}$ and both bodies are moving towards it at speed $\dfrac{v}{2}$. Because of the symmetry, after the collision both must be moving away from the center of mass at the same speed. Adding the speed of the center of mass to both, we find that the body that was moving is now stopped and the other is moving away at speed v. The bodies have exchanged their velocities. Regardless of the velocities of the bodies, a switch to the center of mass frame leads us to the same conclusion. Therefore, the final velocities are given by

$$v_1 = u_2$$
$$v_2 = u_1.$$

In general, when the initial velocities are known, the final velocities are given by

$$v_1 = \left(\frac{m_1 - m_2}{m_1 + m_2} \right) u_1 + \left(\frac{2m_2}{m_1 + m_2} \right) u_2$$

$$v_2 = \left(\frac{m_2 - m_1}{m_1 + m_2} \right) u_2 + \left(\frac{2m_1}{m_1 + m_2} \right) u_1.$$

If one body has much greater mass than the other, its velocity will be little affected by a collision while the other body will experience a large change.

Inelastic Collisions

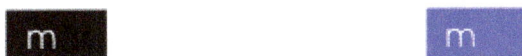

A perfectly inelastic collision between equal masses

In an inelastic collision, some of the kinetic energy of the colliding bodies is converted into other forms of energy (such as heat or sound). Examples include traffic collisions, in which the effect of lost kinetic energy can be seen in the damage to the vehicles; electrons losing some of their energy to atoms (as in the Franck–Hertz experiment); and particle accelerators in which the kinetic energy is converted into mass in the form of new particles.

In a perfectly inelastic collision (such as a bug hitting a windshield), both bodies have the same motion afterwards. If one body is motionless to begin with, the equation for conservation of momentum is

$$m_1 u_1 = (m_1 + m_2) v,$$

so

$$v = \frac{m_1}{m_1 + m_2} u_1.$$

In a frame of reference moving at the speed v), the objects are brought to rest by the collision and 100% of the kinetic energy is converted.

One measure of the inelasticity of the collision is the coefficient of restitution C_R, defined as the ratio of relative velocity of separation to relative velocity of approach. In applying this measure to ball sports, this can be easily measured using the following formula:

$$C_R = \sqrt{\frac{\text{bounce height}}{\text{drop height}}}.$$

The momentum and energy equations also apply to the motions of objects that begin together and then move apart. For example, an explosion is the result of a chain reaction that transforms potential energy stored in chemical, mechanical, or nuclear form into kinetic energy, acoustic energy, and electromagnetic radiation. Rockets also make use of conservation of momentum: propellant is thrust outward, gaining momentum, and an equal and opposite momentum is imparted to the rocket.

Multiple Dimensions

Two-dimensional elastic collision. There is no motion perpendicular to the image,
so only two components are needed to represent the velocities and momenta.
It represents velocities after the collision and add vectorially to get the initial velocity.

Real motion has both direction and velocity and must be represented by a vector. In a coordinate system with x, y, z axes, velocity has components v_x in the x direction, v_y in the y direction, v_z in the z direction. The vector is represented by a boldface symbol:

$$\mathbf{v} = \left(v_x, v_y, v_z \right).$$

Similarly, the momentum is a vector quantity and is represented by a boldface symbol:

$$\mathbf{p} = \left(p_x, p_y, p_z \right).$$

The previous equations, work in vector form if the scalars p and v are replaced by vectors p and **v**. Each vector equation represents three scalar equations. For example,

$$\mathbf{p} = m\mathbf{v}$$

represents three equations:

$$p_x = mv_x$$
$$p_y = mv_y$$
$$p_z = mv_z.$$

The kinetic energy equations are exceptions to the above replacement rule. The equations are still one-dimensional, but each scalar represents the magnitude of the vector, for example,

$$v^2 = v_x^2 + v_y^2 + v_z^2.$$

Each vector equation represents three scalar equations. Often coordinates can be chosen so that only two components are needed, as in the figure. Each component can be obtained separately and the results combined to produce a vector result.

A simple construction involving the center of mass frame can be used to show that if a stationary elastic sphere is struck by a moving sphere, the two will head off at right angles after the collision (as in the figure).

Objects of Variable Mass

The concept of momentum plays a fundamental role in explaining the behavior of variable-mass objects such as a rocket ejecting fuel or a star accreting gas. In analyzing such an object, one treats the object's mass as a function that varies with time: $m(t)$. The momentum of the object at time t is therefore $p(t) = m(t)v(t)$. One might then try to invoke Newton's second law of motion by saying that the external force F on the object is related to its momentum $p(t)$ by $F = \dfrac{dp}{dt}$, but this is incorrect, as is the related expression found by applying the product rule to $\dfrac{d(mv)}{dt}$:

$$F = m(t)\frac{dv}{dt} + v(t)\frac{dm}{dt}. \qquad \text{(incorrect)}$$

This equation does not correctly describe the motion of variable-mass objects. The correct equation is

$$F = m(t)\frac{dv}{dt} - u\frac{dm}{dt},$$

where u is the velocity of the ejected/accreted mass *as seen in the object's rest frame*. This is distinct from v, which is the velocity of the object itself as seen in an inertial frame.

This equation is derived by keeping track of both the momentum of the object as well as the momentum of the ejected/accreted mass (dm). When considered together, the object and the mass (dm) constitute a closed system in which total momentum is conserved.

$$P(t+dt) = (m-dm)(v+dv) + dm(v-u) = mv + mdv - udm = P(t) + mdv - udm :$$

Relativistic Mechanics

Lorentz Invariance

Newtonian physics assumes that absolute time and space exist outside of any observer; this gives rise to the Galilean invariance described earlier. It also results in a prediction that the speed of light can vary from one reference frame to another. This is contrary to observation. In the special theory of relativity, Einstein keeps the postulate that the equations of motion do not depend on the reference frame, but assumes that the speed of light c is invariant. As a result, position and time in two reference frames are related by the Lorentz transformation instead of the Galilean transformation.

Consider, for example, a reference frame moving relative to another at velocity v in the x direction. The Galilean transformation gives the coordinates of the moving frame as

$$t' = t$$
$$x' = x - vt$$

while the Lorentz transformation gives

$$t' = \gamma\left(t - \frac{vx}{c^2}\right)$$
$$x' = \gamma\left(x - vt\right)$$

where γ is the Lorentz factor:

$$\gamma = \frac{1}{\sqrt{1 - v^2/c^2}}.$$

Newton's second law, with mass fixed, is not invariant under a Lorentz transformation. However, it can be made invariant by making the *inertial mass* m of an object a function of velocity:

$$m = \gamma m_0;$$

m_0 is the object's invariant mass.

The modified momentum,

$$\mathbf{p} = \gamma m_0 \mathbf{v},$$

obeys Newton's second law:

$$\mathbf{F} = \frac{d\mathbf{p}}{dt}.$$

Within the domain of classical mechanics, relativistic momentum closely approximates Newtonian momentum: at low velocity, $\gamma m_0 \mathbf{v}$ is approximately equal to $m_0 \mathbf{v}$, the Newtonian expression for momentum.

Four-vector Formulation

In the theory of special relativity, physical quantities are expressed in terms of four-vectors that include time as a fourth coordinate along with the three space coordinates. These vectors are generally represented by capital letters, for example R for position. The expression for the four-momentum depends on how the coordinates are expressed. Time may be given in its normal units or multiplied by the speed of light so that all the components of the four-vector have dimensions of length. If the latter scaling is used, an interval of proper time, τ, defined by

$$c^2 d\tau^2 = c^2 dt^2 - dx^2 - dy^2 - dz^2,$$

is invariant under Lorentz transformations (in this expression and in what follows the (+ − − −) metric signature has been used, different authors use different conventions). Mathematically this invariance can be ensured in one of two ways: by treating the four-vectors as Euclidean vectors and multiplying time by $\sqrt{-1}$; or by keeping time a real quantity and embedding the vectors in a Minkowski space. In a Minkowski space, the scalar product of two four-vectors $\mathbf{U} = (U_0, U_1, U_2, U_3)$ and $\mathbf{V} = (V_0, V_1, V_2, V_3)$ is defined as

$$\mathbf{U} \cdot \mathbf{V} = U_0 V_0 - U_1 V_1 - U_2 V_2 - U_3 V_3.$$

In all the coordinate systems, the (contravariant) relativistic four-velocity is defined by

$$\mathbf{U} \equiv \frac{d\mathbf{R}}{d\tau} = \gamma \frac{d\mathbf{R}}{dt},$$

and the (contravariant) four-momentum is

$$\mathbf{P} = m_0 \mathbf{U},$$

where m_0 is the invariant mass. If $R = (ct,x,y,z)$ (in Minkowski space), then

$$\mathbf{P} = \gamma m_0 (c, \mathbf{v}) = (mc, \mathbf{p}).$$

Using Einstein's mass-energy equivalence, $E = mc^2$, this can be rewritten as

$$\mathbf{P} = \left(\frac{E}{c}, \mathbf{p} \right).$$

Thus, conservation of four-momentum is Lorentz-invariant and implies conservation of both mass and energy.

The 4-Momentum is related to the 4-WaveVector in Special Relativity

$$\mathbf{P} = \left(\frac{E}{c}, \vec{\mathbf{p}} \right) = \hbar \mathbf{K} = \hbar \left(\frac{\omega}{c}, \vec{\mathbf{k}} \right) \text{ which is the full 4-vector version of:}$$

The (temporal component) Planck–Einstein relation $E = \hbar \omega$

The (spatial components) de Broglie matter wave relation $\vec{\mathbf{p}} = \hbar \vec{\mathbf{k}}$

The magnitude of the momentum four-vector is equal to $m_0 c$:

$$\| \mathbf{P} \|^2 = \mathbf{P} \cdot \mathbf{P} = \gamma^2 m_0^2 (c^2 - v^2) = (m_0 c)^2,$$

and is invariant across all reference frames.

The relativistic energy–momentum relationship holds even for massless particles such as photons; by setting $m_0 = 0$ it follows that

$$E = pc.$$

In a game of relativistic "billiards", if a stationary particle is hit by a moving particle in an elastic collision, the paths formed by the two afterwards will form an acute angle. This is unlike the non-relativistic case where they travel at right angles.

Generalized Coordinates

Newton's laws can be difficult to apply to many kinds of motion because the motion is limited by *constraints*. For example, a bead on an abacus is constrained to move along its wire and a pendulum bob is constrained to swing at a fixed distance from the pivot. Many such constraints can be incorporated by changing the normal Cartesian coordinates to a set of *generalized coordinates* that may be fewer in number. Refined mathematical methods have been developed for solving mechanics problems in generalized coordinates. They introduce a *generalized momentum*, also known as the *canonical* or *conjugate momentum*, that extends the concepts of both linear momentum and angular momentum. To distinguish it from generalized momentum, the product of mass and velocity is also referred to as *mechanical, kinetic* or *kinematic momentum*. The two main methods are described below.

Lagrangian Mechanics

In Lagrangian mechanics, a Lagrangian is defined as the difference between the kinetic energy T and the potential energy V:

$$\mathcal{L} = T - V.$$

If the generalized coordinates are represented as a vector q = $(q_1, q_2, \ldots, q_N)$ and time differentiation is represented by a dot over the variable, then the equations of motion (known as the Lagrange or Euler–Lagrange equations) are a set of N equations:

$$\frac{d}{dt}\left(\frac{\partial \mathcal{L}}{\partial \dot{q}_j}\right) - \frac{\partial \mathcal{L}}{\partial q_j} = 0.$$

If a coordinate q_i is not a Cartesian coordinate, the associated generalized momentum component p_i does not necessarily have the dimensions of linear momentum. Even if q_i is a Cartesian coordinate, p_i will not be the same as the mechanical momentum if the potential depends on velocity. Some sources represent the kinematic momentum by the symbol Π.

In this mathematical framework, a generalized momentum is associated with the generalized coordinates. Its components are defined as

$$p_j = \frac{\partial \mathcal{L}}{\partial \dot{q}_j}.$$

Each component p_j is said to be the *conjugate momentum* for the coordinate q_j.

Now if a given coordinate q_i does not appear in the Lagrangian (although its time derivative might appear), then

$$p_j = \text{constant}.$$

This is the generalization of the conservation of momentum.

Even if the generalized coordinates are just the ordinary spatial coordinates, the conjugate momenta are not necessarily the ordinary momentum coordinates. An example is found in the section on electromagnetism.

Hamiltonian Mechanics

In Hamiltonian mechanics, the Lagrangian (a function of generalized coordinates and their derivatives) is replaced by a Hamiltonian that is a function of generalized coordinates and momentum. The Hamiltonian is defined as

$$\mathcal{H}(\mathbf{q},\mathbf{p},t) = \mathbf{p} \cdot \dot{\mathbf{q}} - \mathcal{L}\left(\mathbf{q},\dot{\mathbf{q}},t\right),$$

where the momentum is obtained by differentiating the Lagrangian as above. The Hamiltonian equations of motion are

$$\dot{q}_i = \frac{\partial \mathcal{H}}{\partial p_i}$$

$$-\dot{p}_i = \frac{\partial \mathcal{H}}{\partial q_i}$$

$$-\frac{\partial \mathcal{L}}{\partial t} = \frac{d\mathcal{H}}{dt}.$$

As in Lagrangian mechanics, if a generalized coordinate does not appear in the Hamiltonian, its conjugate momentum component is conserved.

Symmetry and Conservation

Conservation of momentum is a mathematical consequence of the homogeneity (shift symmetry) of space (position in space is the canonical conjugate quantity to momentum). That is, conservation of momentum is a consequence of the fact that the laws of physics do not depend on position; this is a special case of Noether's theorem.

Electromagnetism

In Newtonian mechanics, the law of conservation of momentum can be derived from the law of action and reaction, which states that every force has a reciprocating equal and opposite force. Under some circumstances, moving charged particles can exert forces on each other in non-opposite directions. Moreover, Maxwell's equations, the foundation of classical electrodynamics, are Lorentz-invariant. Nevertheless, the combined momentum of the particles and the electromagnetic field is conserved.

Vacuum

In Maxwell's equations, the forces between particles are mediated by electric and magnetic fields. The electromagnetic force (*Lorentz force*) on a particle with charge q due to a combination of electric field E and magnetic field (as given by the "B-field" B) is

$$F = q(E + v \times B).$$

This force imparts a momentum to the particle, so by Newton's second law the particle must impart a momentum to the electromagnetic fields.

In a vacuum, the momentum per unit volume is

$$g = \frac{1}{\mu_0 c^2} E \times B,$$

where μ_0 is the vacuum permeability and c is the speed of light. The momentum density is proportional to the Poynting vector S which gives the directional rate of energy transfer per unit area:

$$g = \frac{S}{c^2}.$$

If momentum is to be conserved over the volume V over a region Q, changes in the momentum of matter through the Lorentz force must be balanced by changes in the momentum of the electromagnetic field and outflow of momentum. If P_{mech} is the momentum of all the particles in Q, and the particles are treated as a continuum, then Newton's second law gives

$$\frac{dP_{mech}}{dt} = \iiint_Q (\rho E + J \times B) dV.$$

The electromagnetic momentum is

$$P_{field} = \frac{1}{\mu_0 c^2} \iiint_Q E \times B \, dV,$$

and the equation for conservation of each component i of the momentum is

$$\frac{d}{dt}(P_{mech} + P_{field})_i = \iint_\sigma \left(\sum_j T_{ij} n_j \right) d\Sigma.$$

The term on the right is an integral over the surface area Σ of the surface σ representing momentum flow into and out of the volume, and n_j is a component of the surface normal of S. The quantity T_{ij} is called the Maxwell stress tensor, defined as

$$T_{ij} \equiv \epsilon_0 \left(E_i E_j - \frac{1}{2} \delta_{ij} E^2 \right) + \frac{1}{\mu_0} \left(B_i B_j - \frac{1}{2} \delta_{ij} B^2 \right).$$

Media

The above results are for the *microscopic* Maxwell equations, applicable to electromagnetic forces in a vacuum (or on a very small scale in media). It is more difficult to define momentum density in media because the division into electromagnetic and mechanical is arbitrary. The definition of electromagnetic momentum density is modified to

$$\mathbf{g} = \frac{1}{c^2}\mathbf{E}\times\mathbf{H} = \frac{\mathbf{S}}{c^2},$$

where the H-field H is related to the B-field and the magnetization M by

$$\mathbf{B} = \mu_0\left(\mathbf{H}+\mathbf{M}\right).$$

The electromagnetic stress tensor depends on the properties of the media.

Particle in Field

If a charged particle q moves in an electromagnetic field, neither its kinetic momentum $m\mathbf{v}$ nor its canonical momentum is conserved.

Lagrangian and Hamiltonian Formulation

The *kinetic momentum* p is different from the *canonical momentum* P (synonymous with the generalized momentum) conjugate to the ordinary position coordinates r, because P includes a contribution from the electric potential $\varphi(r, t)$ and vector potential $A(r, t)$:

	Classical mechanics	**Relativistic mechanics**
Lagrangian	$\mathcal{L} = \frac{m}{2}\dot{\mathbf{r}}\cdot\dot{\mathbf{r}} + q\mathbf{A}\cdot\dot{\mathbf{r}} - q\varphi$	$\mathcal{L} = -mc^2\gamma^{-1} + q\mathbf{A}\cdot\dot{\mathbf{r}} - q\varphi$
Canonical momentum $\mathbf{P} = \dfrac{\partial L}{\partial\dot{\mathbf{r}}}$	$\mathbf{P} = m\dot{\mathbf{r}} + q\mathbf{A}$	$\mathbf{P} = \gamma m\dot{\mathbf{r}} + q\mathbf{A}$
Kinetic momentum $\mathbf{p} = m\dot{\mathbf{r}}$	$m\dot{\mathbf{r}} = \mathbf{P} - q\mathbf{A}$	$\gamma m\dot{\mathbf{r}} = \mathbf{P} - q\mathbf{A}$
Hamiltonian	$\mathcal{H} = T + V$ $= \dfrac{\mathbf{p}^2}{2m} + V$ $= \dfrac{1}{2m}(\mathbf{P}-q\mathbf{A})^2 + q\varphi$	$\mathcal{H} = \mathbf{P}\cdot\dot{\mathbf{r}} - \mathcal{L}$ $= \gamma mc^2 + q\varphi$ $= \sqrt{c^2(\mathbf{P}-q\mathbf{A})^2 + (mc^2)^2} + q\varphi$

where $\dot{r}$ = v is the velocity, q is the electric charge of the particle

and $\gamma = (1 - \dfrac{\dot{r} \cdot \dot{r}}{c^2})^{-\frac{1}{2}}$ is the Lorentz factor.

If neither φ nor A depends on position, P is conserved.

The classical Hamiltonian H for a particle in any field equals the total energy of the system – the kinetic energy T = $\dfrac{P^2}{2m}$ (where p² = p · p) plus the potential energy

V. For a particle in an electromagnetic field, the potential energy is V = eφ, and since the kinetic energy T always corresponds to the kinetic momentum p, replacing the kinetic momentum by the above equation (p = P − qA) leads to the Hamiltonian in the table.

These Lagrangian and Hamiltonian expressions can derive the Lorentz force.

Canonical Commutation Relations

The kinetic momentum (p above) satisfies the commutation relation:

$$\left[p_j, p_k \right] = \frac{i\hbar e}{c} \varepsilon_{jkl} B_l$$

where: j, k, l are indices labelling vector components, B_l is a component of the magnetic field, and ε_{kjl} is the Levi-Civita symbol, here in 3 dimensions.

Quantum Mechanics

In quantum mechanics, momentum is defined as a self-adjoint operator on the wave function. The Heisenberg uncertainty principle defines limits on how accurately the momentum and position of a single observable system can be known at once. In quantum mechanics, position and momentum are conjugate variables.

For a single particle described in the position basis the momentum operator can be written as

$$\mathbf{p} = \frac{\hbar}{i} \nabla = -i\hbar \nabla,$$

where ∇ is the gradient operator, $\hbar$ is the reduced Planck constant, and i is the imaginary unit. This is a commonly encountered form of the momentum operator, though the momentum operator in other bases can take other forms. For example, in momentum space the momentum operator is represented as

$$\mathbf{p}\psi(p) = p\psi(p),$$

where the operator p acting on a wave function $\psi(p)$ yields that wave function multiplied by the value p, in an analogous fashion to the way that the position operator acting on a wave function $\psi(x)$ yields that wave function multiplied by the value x.

For both massive and massless objects, relativistic momentum is related to the phase constant β by

$$p = \hbar\beta$$

Electromagnetic radiation (including visible light, ultraviolet light, and radio waves) is carried by photons.Even though photons (the particle aspect of light) have no mass, they still carry momentum. This leads to applications such as the solar sail. The calculation of the momentum of light within dielectric media is somewhat controversial.

Deformable Bodies and Fluids

Conservation in a Continuum

In fields such as fluid dynamics and solid mechanics, it is not feasible to follow the motion of individual atoms or molecules. Instead, the materials must be approximated by a continuum in which there is a particle or fluid parcel at each point that is assigned the average of the properties of atoms in a small region nearby. In particular, it has a density ρ and velocity v that depend on time t and position r. The momentum per unit volume is ρv.

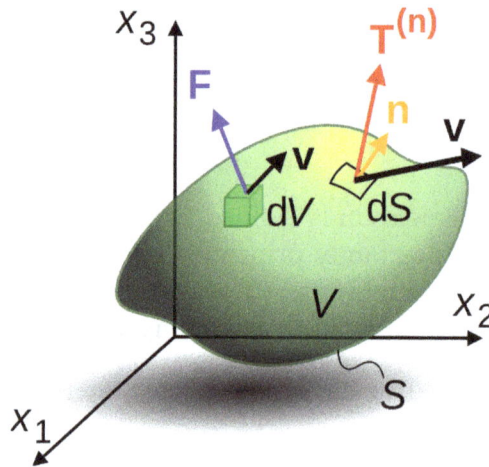

Motion of a material body

Consider a column of water in hydrostatic equilibrium. All the forces on the water are in balance and the water is motionless. On any given drop of water, two forces are balanced. The first is gravity, which acts directly on each atom and molecule inside. The gravitational force per unit volume is ρg, where g is the gravitational acceleration. The second force is the sum of all the forces exerted on its surface by the surrounding water. The force from below is greater than the force from above by just the amount needed to balance gravity. The normal force per unit area is the pressure p. The average force per unit volume inside the droplet is the gradient of the pressure, so the force balance equation is

$$-\nabla p + \rho \mathbf{g} = 0.$$

If the forces are not balanced, the droplet accelerates. This acceleration is not simply the partial derivative $\dfrac{\partial v}{\partial t}$ because the fluid in a given volume changes with time. Instead, the material derivative is needed:

$$\frac{D}{Dt} \equiv \frac{\partial}{\partial t} + \mathbf{v} \cdot \nabla.$$

Applied to any physical quantity, the material derivative includes the rate of change at a point and the changes due to advection as fluid is carried past the point. Per unit volume, the rate of change in momentum is equal to $\rho \dfrac{Dv}{Dt}$. This is equal to the net force on the droplet.

Forces that can change the momentum of a droplet include the gradient of the pressure and gravity, as above. In addition, surface forces can deform the droplet. In the simplest case, a shear stress τ, exerted by a force parallel to the surface of the droplet, is proportional to the rate of deformation or strain rate. Such a shear stress occurs if the fluid has a velocity gradient because the fluid is moving faster on one side than another. If the speed in the x direction varies with z, the tangential force in direction x per unit area normal to the z direction is

$$\sigma_{zx} = -\mu \frac{\partial v_x}{\partial z},$$

where μ is the viscosity. This is also a flux, or flow per unit area, of x-momentum through the surface.

Including the effect of viscosity, the momentum balance equations for the incompressible flow of a Newtonian fluid are

$$\rho \frac{D\mathbf{v}}{Dt} = -\nabla p + \mu \nabla^2 \mathbf{v} + \rho \mathbf{g}.$$

These are known as the Navier–Stokes equations.

The momentum balance equations can be extended to more general materials, including solids. For each surface with normal in direction i and force in direction j, there is a stress component σ_{ij}. The nine components make up the Cauchy stress tensor σ, which includes both pressure and shear. The local conservation of momentum is expressed by the Cauchy momentum equation:

$$\rho \frac{D\mathbf{v}}{Dt} = \nabla \cdot \sigma + \mathbf{f},$$

where f is the body force.

The Cauchy momentum equation is broadly applicable to deformations of solids and liquids. The relationship between the stresses and the strain rate depends on the properties of the material.

Acoustic Waves

A disturbance in a medium gives rise to oscillations, or waves, that propagate away from their source. In a fluid, small changes in pressure p can often be described by the acoustic wave equation:

$$\frac{\partial^2 p}{\partial t^2} = c^2 \nabla^2 p,$$

where c is the speed of sound. In a solid, similar equations can be obtained for propagation of pressure (P-waves) and shear (S-waves).

The flux, or transport per unit area, of a momentum component ρv_j by a velocity v_i is equal to $\rho \, v_j v_j$. In the linear approximation that leads to the above acoustic equation, the time average of this flux is zero. However, nonlinear effects can give rise to a nonzero average. It is possible for momentum flux to occur even though the wave itself does not have a mean momentum.

History of the Concept

In about 530 AD, working in Alexandria, Byzantine philosopher John Philoponus developed a concept of momentum in his commentary to Aristotle's *Physics*. Aristotle claimed that everything that is moving must be kept moving by something. For example, a thrown ball must be kept moving by motions of the air. Most writers continued to accept Aristotle's theory until the time of Galileo, but a few were skeptical. Philoponus pointed out the absurdity in Aristotle's claim that motion of an object is promoted by the same air that is resisting its passage. He proposed instead that an impetus was imparted to the object in the act of throwing it. Ibn Sīnā (also known by his Latinized name Avicenna) read Philoponus and published his own theory of motion in *The Book of Healing* in 1020. He agreed that an impetus is imparted to a projectile by the thrower; but unlike Philoponus, who believed that it was a temporary virtue that would decline even in a vacuum, he viewed it as a persistent, requiring external forces such as air resistance to dissipate it. The work of Philoponus, and possibly that of Ibn Sīnā, was read and refined by the European philosophers Peter Olivi and Jean Buridan. Buridan, who in about 1350 was made rector of the University of Paris, referred to impetus being proportional to the weight times the speed. Moreover, Buridan's theory was different from his predecessor's in that he did not consider impetus to be self-dissipating, asserting that a body would be arrested by the forces of air resistance and gravity which might be opposing its impetus.

René Descartes believed that the total "quantity of motion" in the universe is conserved, where the quantity of motion is understood as the product of size and speed. This should not be read as a statement of the modern law of momentum, since he had no concept of mass as distinct from weight and size, and more importantly he believed that it is speed rather than velocity that is conserved. So for Descartes if a moving object were to bounce off a surface, changing its direction but not its speed, there would be no change in its quantity of motion. Galileo, later, in his *Two New Sciences*, used the Italian word *impeto*.

Leibniz, in his "Discourse on Metaphysics", gave an argument against Descartes' construction of the conservation of the "quantity of motion" using an example of dropping blocks of different sizes different distances. He points out that force is conserved but quantity of motion, construed as the product of size and speed of an object, is not conserved.

The first correct statement of the law of conservation of momentum was by English mathematician John Wallis in his 1670 work, *Mechanica sive De Motu, Tractatus Geometricus*: "the initial state of the body, either of rest or of motion, will persist" and "If the force is greater than the resistance, motion will result". Wallis uses *momentum* and *vis* for force. Newton's *Philosophiæ Naturalis Principia Mathematica*, when it was first published in 1687, showed a similar casting around for words to use for the mathematical momentum. His Definition II defines *quantitas motus*, "quantity of motion", as "arising from the velocity and quantity of matter conjointly", which identifies it as momentum. Thus when in Law II he refers to *mutatio motus*, "change of motion", being proportional to the force impressed, he is generally taken to mean momentum and not motion. It remained only to assign a standard term to the quantity of motion. The first use of "momentum" in its proper mathematical sense is not clear but by the time of Jenning's *Miscellanea* in 1721, five years before the final edition of Newton's *Principia Mathematica*, momentum M or "quantity of motion" was being defined for students as "a rectangle", the product of Q and V, where Q is "quantity of material" and V is "velocity", $\frac{s}{t}$.

Momentum Conservation Equation

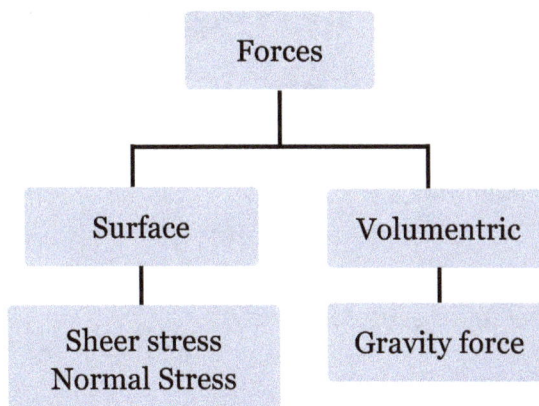

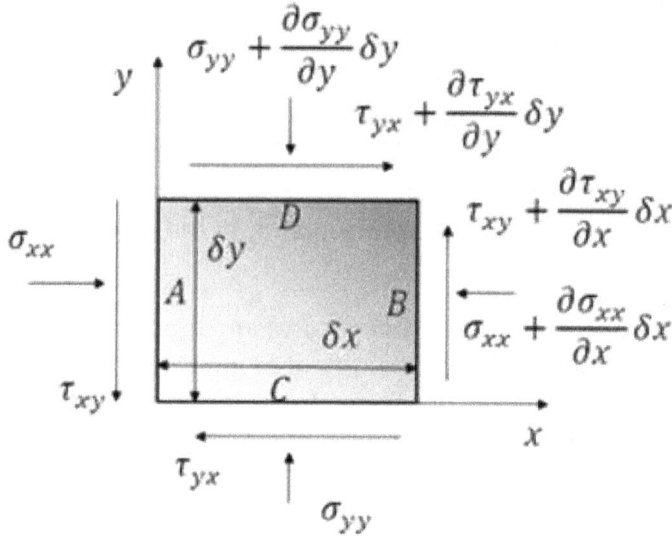

$$\left.\begin{array}{l} \text{Rate of momentum} \\ \text{Accumulation in fluid element} \end{array}\right\} =$$

$$\left.\begin{array}{l} \text{Rate of momentum} \\ \text{into fluid element} \end{array}\right\} - \left\{\begin{array}{l} \text{Rate of momentum} \\ \text{out of fluid element} \end{array}\right\} + \left\{\begin{array}{l} \text{Sums of forces acting} \\ \text{on the system} \end{array}\right\}$$

Rate of momentum accumulation in x-direction = $\dfrac{\partial(\rho V_x)}{\partial t}(\delta x \times \delta y \times 1)$

Rate of momentum accumulation in y-direction = $\dfrac{\partial(\rho V_y)}{\partial t}(\delta x \times \delta y \times 1)$

$$\left.\begin{array}{l} \textit{Momentum in x} - \textit{direction into fluid element} \\ \textit{across face A} \end{array}\right\} = \rho V_x.V_x(\delta y \times 1)$$

$$\left.\begin{array}{l} \textit{Momentum in x} - \textit{direction leaving} \\ \textit{the fluid element across face B} \end{array}\right\} = \rho V_x V_x(\delta y \times 1) + \dfrac{\partial}{\partial x}[\rho V_x V_x(\delta y \times 1)\delta x]$$

$$\left.\begin{array}{l} \textit{Momentum in y} - \textit{direction entering the fluid} \\ \textit{element through face C} \end{array}\right\} = \rho V_x V_y(\delta x \times 1)$$

$$\left.\begin{array}{l} \textit{Momentum in y} - \textit{direction leaving the fluid} \\ \textit{element across face D} \end{array}\right\} = \rho V_x V_y(\delta x \times 1) + \dfrac{\partial}{\partial y}[\rho V_x V_y(\delta x \times 1)\delta y]$$

$$\left(\sigma_{xx} + \frac{\partial \sigma_{xx}}{\partial x}\delta x - \sigma_{xx}\right)(\delta y \times 1)$$

Net forces acting on the fluid element in x-direction $= +\left(\tau_{yx} + \frac{\partial \tau_{yx}}{\partial y}\tau_{yx}\right)(\delta x \times 1)$

$$\Rightarrow \left(\frac{\partial \sigma_{xx}}{\partial x} + \frac{\partial \tau_{yx}}{\partial y}\right)(\delta x \times \delta y)$$

$$\left.\begin{array}{l} \textit{Net body forces acting in fluid element} \\ \textit{in the } x - \textit{direction} \end{array}\right\} = \rho f_x = (\delta x \times \delta y)$$

$$\left.\begin{array}{l} \textit{Momentum equation for fluid element} \\ \textit{in } x - \textit{direction} \end{array}\right\} = \frac{\partial(\rho V_x)}{\partial t} + \frac{\partial(\rho V_x V_x)}{\partial x} + \frac{\partial(\rho V_x V_x)}{\partial y} = \frac{\partial \sigma_{xx}}{\partial x} + \frac{\partial \tau_{yx}}{\partial y} + \rho f_x$$

$$\left.\begin{array}{l} \textit{Momentum equation for fluid} \\ \textit{element in } x - \textit{direction} \end{array}\right\} = \frac{\partial(\rho V_y)}{\partial t} + \frac{\partial(\rho V_x V_y)}{\partial x} + \frac{\partial(\rho V_y V_y)}{\partial y} = \frac{\partial \sigma_{xy}}{\partial x} + \frac{\partial \tau_{yy}}{\partial y} + \rho f_y$$

where, ρV_x , ρV_y are components of mass velocity vector in x and y direction τ and σ are surface stresses

Applying Stokes viscosity law, the surface stresses are given by

$$\tau_{xy} = \tau_{yx} = \mu\left(\frac{\partial V_x}{\partial y} + \frac{\partial V_y}{\partial x}\right)$$

$$\sigma_{xx} = \mu\left(2\frac{\partial V_x}{\partial x} - \frac{2}{3}(\nabla.V)\right) - P \approx -P$$

$$\sigma_{yy} = \mu\left(2\frac{\partial V_y}{\partial y} - \frac{2}{3}(\nabla.V)\right) - P \approx -P$$

$$\left.\begin{array}{l} \textit{Momentum equation for} \\ \textit{fluid element in } x - \textit{direction} \end{array}\right\} = \frac{\partial(\rho V_x)}{\partial t} + \frac{\partial(\rho V_x V_x)}{\partial x} + \frac{\partial(\rho V_x V_y)}{\partial y}$$

$$= -\frac{\partial P}{\partial_x} + \frac{\partial}{\partial x}\left(\mu\frac{\partial V_x}{\partial x}\right) + \frac{\partial}{\partial y}\left(\mu\frac{\partial V_x}{\partial y}\right) + \rho f_x$$

$$\left.\begin{array}{l}\text{Momentum equation for} \\ \text{fluid element in } y - \text{direction}\end{array}\right\} = \frac{\partial(\rho V_y)}{\partial t} + \frac{\partial(\rho V_x V_y)}{\partial x} + \frac{\partial(\rho V_y V_y)}{\partial y}$$

$$= -\frac{\partial P}{\partial y} + \frac{\partial}{\partial x}\left(\mu\frac{\partial V_y}{\partial x}\right) + \frac{\partial}{\partial y}\left(\mu\frac{\partial V_y}{\partial y}\right) + \rho f_y$$

Species Transport Equation

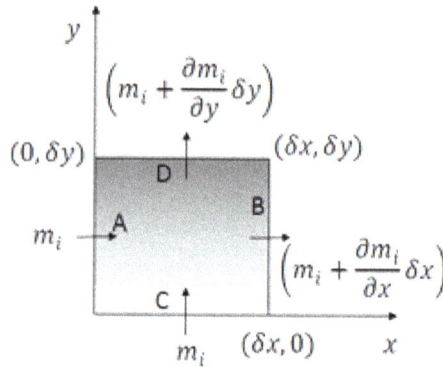

$$\left.\begin{array}{l}\text{Rate of accumulation of mass} \\ \text{of species A in fluid element}\end{array}\right\} = \left\{\begin{array}{l}\text{Rate of mass of species A} \\ \text{into fluid element}\end{array}\right\}$$

$$-\left\{\begin{array}{l}\text{Rate of mass of species A} \\ \text{out of fluid element}\end{array}\right\} + \left\{\begin{array}{l}\text{Mass production rate of species A} \\ \text{due to chemical reaction}\end{array}\right\}$$

Rate of accumulation in fluid element= $\dfrac{\partial(\rho.Y_i.\delta x.\delta y.1)}{\partial t}$

Rate of mass of species A into fluid element across face A = $\dot{m}_i''(\delta y.1)$

$$\left.\begin{array}{l}\text{By Taylor's series expansion, the rate of mass of} \\ \text{species A leaving fluid element across face B}\end{array}\right\} = \left(\dot{m}_i'' + \frac{\partial(\dot{m}_i'')}{\partial x}\delta x\right)(\delta y \times 1)$$

Net efflux in x direction = $\dfrac{\partial \dot{m}_i''}{\partial x} \times (\delta x \times \delta y)$

Net efflux in y direction = $\dfrac{\partial \dot{m}_i''}{\partial y} \times (\delta x \times \delta y)$

Mass production rate of ith species due to chemical reaction = $\dot{m}_i''' \times (\delta x \times \delta y \times 1)$

According to Fick's law, $\dot{m}_i'' = Y_i \sum \dot{m}_i'' - \rho D \left(\dfrac{\partial Y_i}{\partial x} \right)$

Species transport equation is given by,

$$\frac{\partial(\rho Y_i)}{\partial t} + \frac{\partial(\rho V_x V_i)}{\partial x} + \frac{\partial(\rho V_y V_i)}{\partial y} = \frac{\partial}{\partial x}\left(\rho D \frac{\partial Y_i}{\partial x} \right) + \frac{\partial}{\partial y}\left(\rho D \frac{\partial Y_i}{\partial y} \right) + \dot{m}_i''$$

Energy Transport Equation

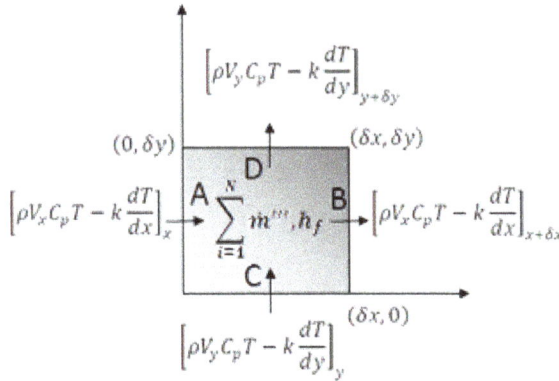

$$\left\{ \begin{array}{l} \text{Rate of energy accumulation} \\ \text{in fluid element} \end{array} \right\} = \left\{ \begin{array}{l} \text{Rate of energy into} \\ \text{fluid element} \end{array} \right\}$$

$$- \left\{ \begin{array}{l} \text{Rate of energy out of fluid} \\ \text{element} \end{array} \right\} + \left\{ \begin{array}{l} \text{Rate of heat added by} \\ \text{chemical reactions} \end{array} \right\}$$

Heat accumulated in the fluid element $= \dfrac{\partial}{\partial t}[\rho C_p T (\delta x \times \delta y \times 1)]$

Amount of heat entering into fluid element $(\rho V_x \times C_p T \times \delta y \times 1) + \left(-k \dfrac{\partial T}{\partial x} \right)(\delta y \times 1)$

through face 'A' is given by:

Amount of heat leaving from the fluid $[(\rho V_x \times C_p T \times \delta x \times 1) + \left(-k \dfrac{\partial T}{\partial x} \right)(\delta y \times 1)]$

element through face 'B' is given by:

$$+ \left[\frac{\partial}{\partial x}(\rho V_x \times C_p T \times \delta y \times 1) + \left(-k \frac{\partial T}{\partial x} \right)(\delta y \times 1) \right](\delta x)$$

Net efflux in x direction $\dfrac{\partial}{\partial x}\left[(\rho V_x \times C_p T \times \delta y \times 1) + \left(-k \dfrac{\partial T}{\partial x} \right)(\delta y \times 1) \right](\delta x)$

Amount of heat interaction in y-direction through faces 'C' and 'D' is

$$\frac{\partial}{\partial x}\left[(-\rho V_x \times C_p T \times \delta y \times 1) + \left(k\frac{\partial T}{\partial x}\right)(\delta y \times 1)\right](\delta x)$$

In a fluid element, heat may be absorbed or removed due to chemical reaction

$$F + Ox \rightarrow P$$

The amount of heat interaction in fluid element per unit area:

$$\sum_i^N \dot{m}_i'' h_{f,i}^0 (\delta x \times \delta y)$$

By striking out an energy balance, the energy equation for a multi-component reactive system because.

$$\frac{\partial}{\partial t}(\rho C_p T) + \frac{\partial}{\partial x}(\rho V_x C_p T) + \frac{\partial}{\partial y}(\rho V_y C_p T) = \frac{\partial}{\partial x}\left(k\frac{\partial T}{\partial x}\right) + \frac{\partial}{\partial y}\left(k\frac{\partial T}{\partial y}\right) - \sum_i^N \dot{m}_i'' h_{f,i}^0$$

Boundary Layer Concept

- Velocity of fluid increases from zero at wall to free stream velocity

- Velocity gradients appear near a thin region adjacent to wall

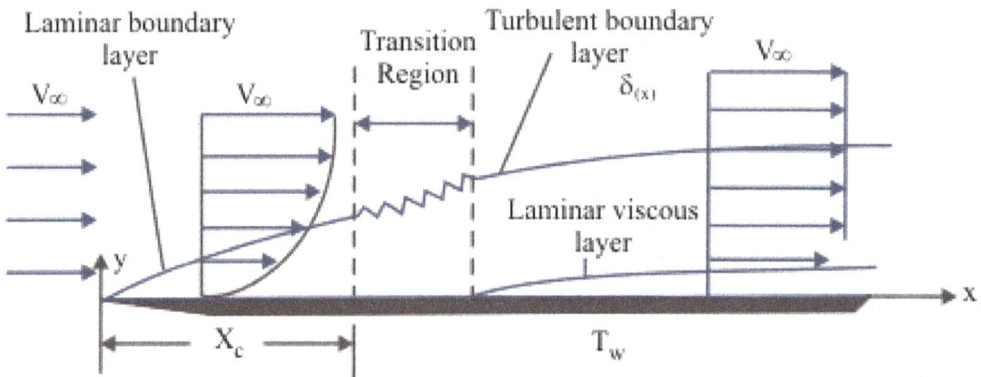

Flow of viscous fluid over a flat plate

- The thin region adjacent to wall surface is the boundary layer

- Wall friction-causes reduction in velocity near the wall

- Boundary layer thickness (δ) = 0.99 times the free stream velocity V_∞

Transport in Turbulent Flow

Turbulent Flow

- At high Reynolds and Grashof's number, the properties, velocity and temperature exhibits random variation.

- Eddies move randomly back and forth across the adjacent fluid layers.

- Turbulence reduces the B.L. thickness.

- Enhanced mass, momentum, and energy transfer rates.

$$V_x = \overline{V}_x + V_x'$$

$$V_y = \overline{V}_y + V_y'$$

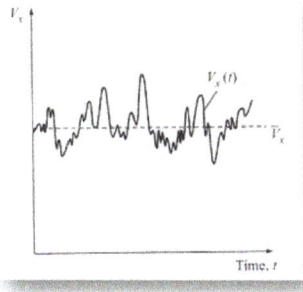

Where , $\overline{V}$ – Time averaged value of velocity

V' – Fluctuating component of velocity

Turbulent diffusivity is given by,

$$\tau_T = \rho V_T \frac{d\overline{V}_x}{dy}; \quad \dot{q}_T'' = -\rho C_p a_T \frac{d\overline{T}}{dy};$$

$$\dot{m}_{AT}'' = -\rho D_T \frac{d\overline{Y}_A}{dy}$$

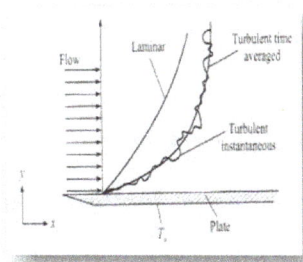

Characterization of Turbulent Flow

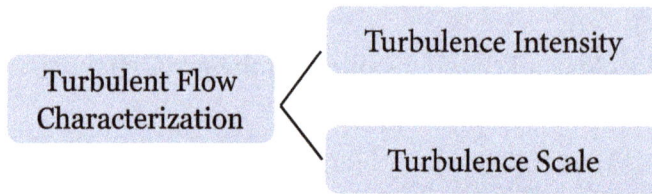

Length Scale of Turbulence

- The distance covered by an eddy before it disappears or loses its identity.

Intensity of Turbulence

- Measure of violence of eddies.

Turbulence Intensity

$$I = \frac{\sqrt{(V_x'^2 + V_y'^2 + V_z'^2)/3}}{\bar{V}}$$

Length Scales used in Turbulent Flow

1. Macroscopic scale, L (Characteristic width of flow)

2. Integral Scale, l_0

3. Taylor micro scale, l_λ

4. Kolmogorov length Scale, l_k

Taylor microscale, $l_\lambda = \dfrac{V_{x,rms}'}{\left[\overline{\left(\dfrac{\partial V_x}{\partial x}\right)^2}\right]^{0.5}}$

where, $\overline{\left(\dfrac{\partial V_x}{\partial x}\right)^2}$ is the mean strain rate

Kolmogorov length scale, $l_K = \left[\dfrac{2\nu^3 l_0}{3V_{rms}'^3}\right]^{1/4}$

Note: Kolmogorov length scale (l_k) is related to integral length scale (l_0)

(l_K) - Thickness of the smallest vortex present in turbulent flow

Turbulent Reynolds number based on the length scales

$$\text{Re}_L = \frac{V'_{rms}L}{\nu} \quad \text{Re}_{l_0} = \frac{V'_{rms}l_0}{\nu} \quad \text{Re}_{l_\lambda} = \frac{V'_{rms}l_\lambda}{\nu} \quad \text{Re}_{lK} = \frac{V'_{rms}l_K}{\nu}$$

Note: V'_{rms} is the characteristic velocity

Turbulent Boundary layer

Consider 2D steady incompressible turbulent flow over a flat plate,

Momentum equation in x direction is given by,

$$\rho\left(\overline{V}_x \frac{\partial \overline{V}_x}{\partial x} + \overline{V}_y \frac{\partial \overline{V}_x}{\partial y}\right) = -\frac{\partial \overline{P}}{\partial x} + \frac{\partial}{\partial y}\left(\mu \frac{\partial \overline{V}_x}{\partial y} - \rho \overline{V'_x V'_y}\right)$$

The term $\rho \overline{V'_x V'_y}$ is known as Reynolds stress

Energy equation for turbulent boundary layer is given by,

$$\rho C_p\left(\overline{V}_x \frac{\partial \overline{T}}{\partial x} + \overline{V}_y \frac{\partial \overline{T}}{\partial y}\right) = \frac{\partial}{\partial y}\left(k \frac{\partial \overline{T}}{\partial y} - \rho C_p \overline{V'_x T}\right)$$

A simple model for Reynolds stress suggested by Bossinesq,

Similarly,

$$-\rho \overline{V'_x V'_y} = \rho \nu_T \frac{\partial \overline{V}_x}{\partial y} \text{ ; where, } \nu_T \text{ is the turbulent diffusivity}$$

$$-\overline{V'_x T} = \propto_T \frac{\partial \overline{T}}{\partial y} \text{ ; where, } \propto_T \text{ is the eddy diffusivity}$$

In analogy to kinetic theory of gases, Prandtl suggested an expression for turbulent diffusivity

$$\nu_T = l_m T$$

Where, l_m is the mixing length, and I is the turbulence intensity

$$I \propto l_m \frac{\partial \overline{V}_x}{\partial y}$$

Combining these two equations,

$$\nu_T = C l_m^2 \frac{d\overline{V}_x}{dy}$$

C, is the constant, obtained from the experimental data

References

- Plawsky, Joel L. (April 2001). Transport phenomena fundamentals (Chemical Industries Series). CRC Press. pp. 1, 2, 3. ISBN 978-0-8247-0500-8

- Z.Y.Wang (2016). "Generalized momentum equation of quantum mechanics". Optical and Quantum Electronics. 48 (2): 1–9. doi:10.1007/s11082-015-0261-8

- Ho-Kim, Quang; Kumar, Narendra; Lam, Harry C. S. (2004). Invitation to Contemporary Physics (illustrated ed.). World Scientific. p. 19. ISBN 978-981-238-303-7

- Barnett, Stephen M. (2010). "Resolution of the Abraham-Minkowski Dilemma". Physical Review Letters. 104 (7). Bibcode:2010PhRvL.104g0401B. PMID 20366861. doi:10.1103/PhysRevLett.104.070401

- Fick, A. (1855). "On liquid diffusion". Poggendorffs Annalen. 94: 59. – reprinted in "On liquid diffusion". Journal of Membrane Science. 100: 33–38. 1995. doi:10.1016/0376-7388(94)00230-v

- Hand, Louis N.; Finch, Janet D. (1998). Analytical mechanics (7th print ed.). Cambridge, England: Cambridge University Press. Chapter 4. ISBN 9780521575720

- McIntyre, M. E. (1981). "On the 'wave momentum' myth". J. Fluid. Mech. 106: 331–347. Bibcode:1981JFM...106..331M. doi:10.1017/s0022112081001626

- G. W. Leibniz (1989). "Discourse on Metaphysics". In Roger Ariew; Daniel Garber. Philosophical Essays. Indianapolis, IN: Hackett Publishing Company, Inc. pp. 49–51. ISBN 0-87220-062-0

- Aydin Sayili (1987). "Ibn Sīnā and Buridan on the Motion of the Projectile". Annals of the New York Academy of Sciences. 500 (1): 477–482. Bibcode:1987NYASA.500..477S. doi:10.1111/j.1749-6632.1987.tb37219.x

- Rescigno, Aldo (2003). Foundation of Pharmacokinetics. New York: Kluwer Academic/Plenum Publishers. p. 19. ISBN 0306477041

4

Chemical Kinetics of Combustion

Many varied reactions take place in combustion. Chemical reaction can be categories in two ways–based on physical state of species and on the basis of reaction rate. The former includes homogenous reaction and heterogeneous reaction, and the latter into explosive and non-explosive. The reaction is dependent on the concentration of species, pressure and temperature. This chapter is an overview of the subject matter incorporating all the major aspects of combustion.

Chemical Kinetics

The specialized branch of physical chemistry dealing with the study of chemical reactions and their governing factors.

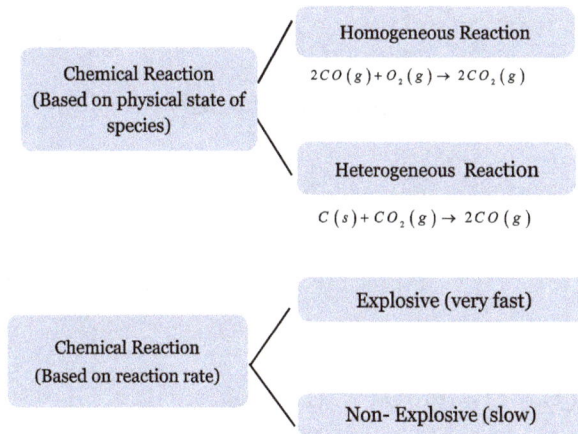

```
                                    ┌─ Homogeneous Reaction
                                    │   $2CO(g) + O_2(g) \rightarrow 2CO_2(g)$
  Chemical Reaction                 │
  (Based on physical state of ──────┤
  species)                          │
                                    └─ Heterogeneous Reaction
                                        $C(s) + CO_2(g) \rightarrow 2CO(g)$


                                    ┌─ Explosive (very fast)
  Chemical Reaction                 │
  (Based on reaction rate) ─────────┤
                                    └─ Non- Explosive (slow)
```

Basic Reaction Kinetics

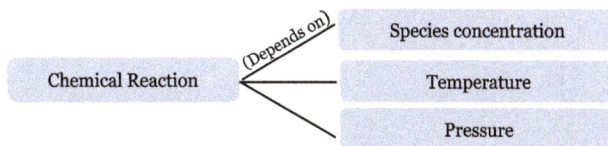

```
                              (Depends on)  ┌─ Species concentration
  Chemical Reaction ──────────────────────  ├─ Temperature
                                            └─ Pressure
```

Reaction Rate

Rate of decrease of reactant concentration or rate of increase of product concentration.

Expressed in terms of mole/m$_3$s

Compact expression for chemical reaction:

$$\sum_{i=1}^{N} v_i' M_i \rightarrow \sum_{i=1}^{N} v_i'' M_i$$

Where,

v_i' and v_i'' are stoichiometric coefficients of reactants and products.

N is the total number of species

M is the arbitrary specification of all chemical species

Expressing the reaction using index notation

$$3H \rightarrow H_2 + H \qquad\qquad H + O_2 \rightarrow HO_2$$

Here, N=2   Here, N=3

$M_1 = H$ $M_2 = H_2$ $M_1 = H$ $M_2 = O_2$ $M_3 = HO_2$

$v_1' = 3$ $v_2' = 0$ $v_1' = 1$ $v_2' = 0$ $v_3' = 0$

$v_1'' = 1$ $v_2'' = 1$ $v_1'' = 0$ $v_2'' = 0$ $v_3'' = 1$

Note: The above reactions are elementary in nature

Global reactions,

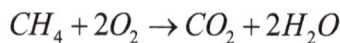

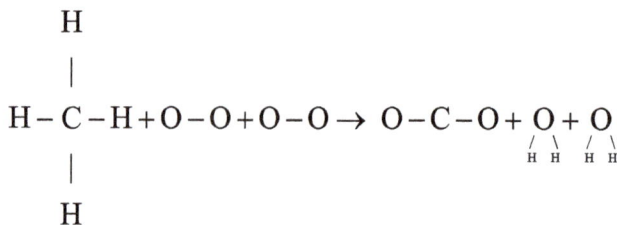

$$2H_2 + O_2 \rightarrow 2H_2O$$

3 bonds have to be broken,

4 bonds have to be formed

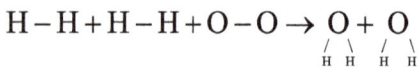

$$H-H+H-H+O-O \rightarrow \underset{\overset{/ \ \ \backslash}{H \ \ H}}{O} + \underset{\overset{/ \ \ \backslash}{H \ \ H}}{O}$$

Global Reactions

$$CH_4 + 2O_2 \rightarrow CO_2 + 2H_2O$$

$$\begin{array}{c} H \\ | \\ H-C-H+O-O+O-O \rightarrow O-C-O+ \underset{\overset{/ \ \backslash}{H \ H}}{O} + \underset{\overset{/ \ \backslash}{H \ H}}{O} \\ | \\ H \end{array}$$

Bimolecular Reactions

Reaction between two molecules,

$$aA + bB \rightarrow cC + dD$$

$$RR = -\frac{1}{a}\frac{dC_A}{dt} = -\frac{1}{b}\frac{dC_B}{dt} = -\frac{1}{c}\frac{dC_c}{dt} = \frac{1}{d}\frac{dC_D}{dt}$$

$$H + O \rightarrow OH \qquad\qquad RR_H = \frac{dC_H}{dt} \propto C_H C_o$$

Law of Mass Action

In chemistry, the law of mass action is the proposition that the rate of a chemical re-action is directly proportional to the product of the activities or concentrations of the reactants. It explains and predicts behaviors of solutions in dynamic equilibrium. Specifically, it implies that for a chemical reaction mixture that is in equilibrium, the ratio between the concentration of reactants and products is constant.

Two aspects are involved in the initial formulation of the law: 1) the equilibrium aspect, concerning the composition of a reaction mixture at equilibrium and 2) the kinetic aspect concerning the rate equations for elementary reactions. Both aspects stem from the research performed by Cato M. Guldberg and Peter Waage between 1864 and 1879 in which equilibrium constants were derived by using kinetic data and the rate equation which they had proposed. Guldberg and Waage also recognized that chemical equilibrium is a dynamic process in which rates of reaction for the forward and backward reactions must be equal at chemical equilibrium. In order to derive the expression of the equilibrium constant appealing to kinetics, the expression of the rate equation must be used. The expression of the rate equations was rediscovered later independently by Jacobus Henricus van't Hoff.

The law is a statement about equilibrium and gives an expression for the equilibrium constant, a quantity characterizing chemical equilibrium. In modern chemistry this is derived using equilibrium thermodynamics.

History

Two chemists generally expressed the composition of a mixture in terms of numer-ical values relating the amount of the product to describe the equilibrium state. Cato Maximilian Guldberg and Peter Waage, building on Claude Louis Berthollet's ideas about reversible chemical reactions, proposed the law of mass action in 1864. These papers, in Norwegian, went largely unnoticed, as did the later publication (in French) of 1867 which contained a modified law and the experimental data on which that law was based.

In 1877 van 't Hoff independently came to similar conclusions, but was unaware of the earlier work, which prompted Guldberg and Waage to give a fuller and further developed account of their work, in German, in 1879. Van 't Hoff then accepted their priority.

1864

The Equilibrium State (Composition)

In their first paper, Guldberg and Waage suggested that in a reaction such as

$$A + B \rightleftharpoons A' + B'$$

the "chemical affinity" or "reaction force" between A and B did not just depend on the chemical nature of the reactants, as had previously been supposed, but also depended on the amount of each reactant in a reaction mixture. Thus the Law of Mass Action was first stated as follows:

> When two reactants, A and B, react together at a given temperature in a "substitution reaction," the affinity, or chemical force between them, is proportional to the active masses, [A] and [B], each raised to a particular power

$$\text{affinity} = \alpha[A]^a[B]^b.$$

In this context a substitution reaction was one such as alcohol + acid $\rightleftharpoons$ ester + water. Active mass was defined in the 1879 paper as "the amount of substance in the sphere of action". For species in solution active mass is equal to concentration. For solids, active mass is taken as a constant. α, a and b were regarded as empirical constants, to be determined by experiment.

At equilibrium, the chemical force driving the forward reaction must be equal to the chemical force driving the reverse reaction. Writing the initial active masses of A,B, A' and B' as p, q, p' and q' and the dissociated active mass at equilibrium as ξ, this equality is represented by

$$\alpha(p - \xi)^a (q - \xi)^b = \alpha'(p' + \xi)^{a'} (q' + \xi)^{b'}$$

ξ represents the amount of reagents A and B that has been converted into A' and B'. Calculations based on this equation are reported in the second paper.

Dynamic Approach to the Equilibrium State

The third paper of 1864 was concerned with the kinetics of the same equilibrium system. Writing the dissociated active mass at some point in time as x, the rate of reaction was given as

$$\left(\frac{dx}{dt}\right)_{forward} = k(p-x)^a(q-x)^b$$

Likewise the reverse reaction of A' with B' proceeded at a rate given by

$$\left(\frac{dx}{dt}\right)_{reverse} = k'(p'+x)^{a'}(q'+x)^{b'}$$

The overall rate of conversion is the difference between these rates, so at equilibrium (when the composition stops changing) the two rates of reaction must be equal. Hence

$$(p-x)^a(q-x)^b = \frac{k'}{k}(p'+x)^{a'}(q'+x)^{b'}$$

1867

The rate expressions given in the 1864 paper could not be integrated, so they were simplified as follows. The chemical force was assumed to be directly proportional to the product of the active masses of the reactants.

$$\text{affinity} = \alpha[A][B]$$

This is equivalent to setting the exponents a and b of the earlier theory to one. The proportionality constant was called an affinity constant, k. The equilibrium condition for an "ideal" reaction was thus given the simplified form

$$k[A]_{eq}[B]_{eq} = k'[A']_{eq}[B']_{eq}$$

$[A]_{eq}$, $[B]_{eq}$ etc. are the active masses at equilibrium. In terms of the initial amounts reagents p,q etc. this becomes

$$(p-\xi)(q-\xi) = \frac{k'}{k}(p'+\xi)(q'+\xi)$$

The ratio of the affinity coefficients, k'/k, can be recognized as an equilibrium constant. Turning to the kinetic aspect, it was suggested that the velocity of reaction, v, is proportional to the sum of chemical affinities (forces). In its simplest form this results in the expression

$$v = \psi(k(p-x)(q-x) - k'(p'+x)(q'+x))$$

where ψ is the proportionality constant. Actually, Guldberg and Waage used a more complicated expression which allowed for interaction between A and A', etc. By making certain simplifying approximations to those more complicated expressions, the rate equation could be integrated and hence the equilibrium quantity ξ could be calculated. The extensive calculations in the 1867 paper gave support to the simplified concept, namely,

> The rate of a reaction is proportional to the product of the active masses of the reagents involved.

This is an alternative statement of the Law of Mass Action.

1879

In the 1879 paper the assumption that reaction rate was proportional to the product of concentrations was justified microscopically in terms of collision theory, as had been developed for gas reactions. It was also proposed that the original theory of the equilibrium condition could be generalised to apply to any arbitrary chemical equilibrium.

$$\text{affinity} = k[A]^\alpha [B]^\beta$$

The exponents α, β etc. are explicitly identified for the first time as the stoichiometric coefficients for the reaction.

Contemporary Statement of the Law

The affinity constants, k_+ and k_-, of the 1879 paper can now be recognised as rate constants. The equilibrium constant, K, was derived by setting the rates of forward and backward reactions to be equal. This also meant that the chemical affinities for the forward and backward reactions are equal. The resultant expression

$$K = \frac{[S]^\sigma [T]^\tau}{[A]^\alpha [B]^\beta}$$

is correct even from the modern perspective, apart from the use of concentrations instead of activities (the concept of chemical activity was developed by Josiah Willard Gibbs, in the 1870s, but was not widely known in Europe until the 1890s). The derivation from the reaction rate expressions is no longer considered to be valid. Nevertheless, Guldberg and Waage were on the right track when they suggested that the driving force for both forward and backward reactions is equal when the mixture is at equilibrium. The term they used for this force was chemical affinity. Today the expression for the equilibrium constant is derived by setting the chemical potential of forward and backward reactions to be equal. The generalisation of the Law of Mass Action, in terms of affinity, to equilibria of arbitrary stoichiometry was a bold and correct conjecture.

The hypothesis that reaction rate is proportional to reactant concentrations is, strictly speaking, only true for elementary reactions (reactions with a single mechanistic step), but the empirical rate expression

$$r_f = k_f[A][B]$$

is also applicable to second order reactions that may not be concerted reactions. Guldberg and Waage were fortunate in that reactions such as ester formation and hydrolysis, on which they originally based their theory, do indeed follow this rate expression.

In general many reactions occur with the formation of reactive intermediates, and/or through parallel reaction pathways. However, all reactions can be represented as a series of elementary reactions and, if the mechanism is known in detail, the rate equation for each individual step is given by the r_f expression so that the overall rate equation can be derived from the individual steps. When this is done the equilibrium constant is obtained correctly from the rate equations for forward and backward reaction rates.

In biochemistry, there has been significant interest in the appropriate mathematical model for chemical reactions occurring in the intracellular medium. This is in contrast to the initial work done on chemical kinetics, which was in simplified systems where reactants were in a relatively dilute, pH-buffered, aqueous solution. In more complex environments, where bound particles may be prevented from disassociation by their surroundings, or diffusion is slow or anomalous, the model of mass action does not always describe the behavior of the reaction kinetics accurately. Several attempts have been made to modify the mass action model, but consensus has yet to be reached. Popular modifications replace the rate constants with functions of time and concentration. As an alternative to these mathematical constructs, one school of thought is that the mass action model can be valid in intracellular environments under certain conditions, but with different rates than would be found in a dilute, simple environment.

The fact that Guldberg and Waage developed their concepts in steps from 1864 to 1867 and 1879 has resulted in much confusion in the literature as to which equation the Law of Mass Action refers. It has been a source of some textbook errors. Thus, today the "law of mass action" sometimes refers to the (correct) equilibrium constant formula, and at other times to the (usually incorrect) r_f rate formula.

Applications to Other Fields

In Semiconductor Physics

The law of mass action also has implications in semiconductor physics. Regardless of doping, the product of electron and hole densities is a constant at equilibrium. This constant depends on the thermal energy of the system (i.e. the product of the Boltzmann constant, k_B, and temperature, T), as well as the band gap (the energy separation between conduction and valence bands, $E_g \equiv E_C - E_V$) and effective density of states in the valence $(N_V(T))$ and conduction $(N_C(T))$ bands. When the equilibrium electron (n_o) and hole (p_o) densities are equal, their density is called the intrinsic carrier density (n_i) as this would be the value of n_o and p_o in a perfect crystal. Note that the final product is independent of the Fermi level (E_F):

$$n_o p_o = \left(N_C e^{-\frac{E_C - E_F}{k_B T}} \right) \left(N_V e^{-\frac{E_F - E_V}{k_B T}} \right) = N_C N_V e^{-\frac{E_g}{k_B T}} = n_i^2$$

Diffusion in Condensed Matter

Yakov Frenkel represented diffusion process in condensed matter as an ensemble of elementary jumps and quasichemical interactions of particles and defects. Henry Eyring applied his theory of absolute reaction rates to this quasichemical representation of diffusion. Mass action law for diffusion leads to various nonlinear versions of Fick's law.

In Mathematical Ecology

The Lotka–Volterra equations describe dynamics of the predator-prey systems. The rate of predation upon the prey is assumed to be proportional to the rate at which the predators and the prey meet; this rate is evaluated as xy, where x is the number of prey, y is the number of predator. This is a typical example of the law of mass action.

In Mathematical Epidemiology

The law of mass action forms the basis of the compartmental model of disease spread in mathematical epidemiology, in which a population of humans, animals or other individuals is divided into categories of susceptible, infected, and recovered (immune). The SIR model is a useful abstraction of disease dynamics which applies well to many disease systems and provides useful outcomes in many circumstances when the Mass Action Principle applies. Individuals in human or animal populations - unlike molecules in an ideal solution - do not mix homogeneously. There are some disease examples in which this non-homogeneity is great enough such that the outputs of the SIR model are invalid. For these situations in which the assumptions of mass action do not apply, more sophisticated graph theory models may be useful.

In Sociophysics

Sociophysics uses tools and concepts from physics and physical chemistry to describe some aspects of social and political behavior. It attempts to explain why and how humans behave much like atoms, at least in some aspects of their collective lives. The law of mass action (generalized if it is necessary) is the main tool to produce the equation of interactions of humans in sociophysics.

Law of Mass Action

The rate of reaction, RR of a chemical species is proportional to the product of the concentrations of the participating chemical species, where each concentration is raised to the power equal to the corresponding stoichiometric coefficient in the chemical reaction.

$$RR_i \propto \prod_{i=1}^{N} C_{M_i}^{v_i}$$

$$RR_i = k \prod_{i=1}^{N} C_{M_i}^{v_i}$$

Where, k is the specific reaction rate or rate coefficient.

Note: k - depends on temperature and activation energy and not on concentration. Law of mass action holds good only for elementary reactions.

Collision Theory

Low concentration = Few collisions   High concentration = More collisions

Reaction rate tends to increase with concentration phenomenon explained by collision theory

Collision theory is a theory proposed independently by Max Trautz in 1916 and William Lewis in 1918, that qualitatively explains how chemical reactions occur and why reaction rates differ for different reactions. The collision theory states that when suitable particles of the reactant hit each other, only a certain percentage of the collisions cause any noticeable or significant chemical change; these successful changes are called successful collisions. The successful collisions have enough energy, also known as activation energy, at the moment of impact to break the preexisting bonds and form all new bonds. This results in the products of the reaction. Increasing the concentration of the reactant particles or raising the temperature, thus bringing about more collisions and therefore many more successful collisions, increases the rate of reaction.

When a catalyst is involved in the collision between the reactant molecules, less energy is required for the chemical change to take place, and hence more collisions have sufficient energy for reaction to occur. The reaction rate therefore increases.

Collision theory is closely related to chemical kinetics.

Rate Constant

The rate constant for a bimolecular gas phase reaction, as predicted by collision theory is:

$$k(T) = Zo\rho \exp\left(\frac{-E_a}{RT}\right)$$

where:

- Zo is the size of the particle .
- ρ is the steric factor.
- E_a is the activation energy of the reaction.
- T is the temperature.
- R is the gas constant.

The collision frequency is:

$$Z = N_A \sigma_{AB} \sqrt{\frac{8k_B T}{\pi \mu_{AB}}}$$

where:

- N_A is the Avogadro constant
- σ_{AB} is the reaction cross section
- k_B is the Boltzmann's constant
- μ_{AB} is the reduced mass of the reactants.

Quantitative Insights

Derivation

Consider the reaction:

$$A + B \rightarrow C$$

In collision theory it is considered that two particles A and B will collide if their nuclei get closer than a certain distance. The area around a molecule A in which it can collide with an approaching B molecule is called the cross section (σ_{AB}) of the reaction and is, in simplified terms, the area corresponding to a circle whose radius

(r_{AB}) is the sum of the radii of both reacting molecules, which are supposed to be spherical. A moving molecule will therefore sweep a volume $\pi r_{AB}^2 c_A$ per second as it moves, where c_A is the average velocity of the particle. (This solely represents the classical notion of a collision of solid balls. As molecules are quantum-mechanical many-particle systems of electrons and nuclei based upon the Coulomb and exchange interactions, generally they neither obey rotational symmetry nor do they have a box potential. Therefore, more generally the cross section is defined as the reaction probability of a ray of A particles per areal density of B targets, which makes the definition independent from the nature of the interaction between A and B. Consequently, the radius r_{AB} can *only* be interpreted as a rough estimate for the "size" of the molecule's wave function.)

From kinetic theory it is known that a molecule of A has an average velocity (different from root mean square velocity) of $c_A = \sqrt{\dfrac{8k_B T}{\pi m_A}}$, where k_B is Boltzmann constant and m_A is the mass of the molecule.

The solution of the two body problem states that two different moving bodies can be treated as one body which has the reduced mass of both and moves with the velocity of the center of mass, so, in this system μ_{AB} must be used instead of m_A.

Therefore, the total collision frequency, of all A molecules, with all B molecules, is:

$$N_A \sigma_{AB} \sqrt{\frac{8k_B T}{\pi \mu_{AB}}}[A][B] = N_A\, r_{AB}^2 \sqrt{\frac{8k_B T}{\pi \mu_{AB}}}[A][B] = Z[A][B]$$

From Maxwell Boltzmann distribution it can be deduced that the fraction of collisions with more energy than the activation energy is $e^{\frac{-E_a}{k_B T}}$. Therefore, the rate of a bimolecular reaction for ideal gases will be:

$$r = Z\rho[A][B]\exp\left(\frac{-E_a}{RT}\right)$$

Where:

- Z is the collision frequency.
- ρ is the steric factor, which will be discussed in detail in the next section.
- E_a is the activation energy of the reaction.
- T is the absolute temperature.

- R is the gas constant.

The product $Z\rho$ is equivalent to the preexponential factor of the Arrhenius equation.

Validity of the Theory and Steric Factor

Once a theory is formulated, its validity must be tested, that is, compare its predictions with the results of the experiments.

When the expression form of the rate constant is compared with the rate equation for an elementary bimolecular reaction, $r = k(T)[A][B]$, it is noticed that

$$k(T) = N_A^2 \sigma_{AB} \sqrt{\frac{8k_B T}{\pi m_A}} \exp\left(\frac{-E_a}{RT}\right)$$

That expression is similar to the Arrhenius equation, and gives the first theoretical explanation for the Arrhenius equation on a molecular basis. The weak temperature dependence of the preexponential factor is so small compared to the exponential factor that it cannot be measured experimentally, that is, *"it is not feasible to establish, on the basis of temperature studies of the rate constant, whether the predicted $T^{\frac{1}{2}}$ dependence of the preexponential factor is observed experimentally"*.

Steric Factor

If the values of the predicted rate constants are compared with the values of known rate constants it is noticed that collision theory fails to estimate the constants correctly and the more complex the molecules are, the more it fails. The reason for this is that particles have been supposed to be spherical and able to react in all directions; that is not true, as the orientation of the collisions is not always the right one. For example, in the hydrogenation reaction of ethylene the H_2 molecule must approach the bonding zone between the atoms, and only a few of all the possible collisions fulfill this requirement.

To alleviate this problem, a new concept must be introduced: the steric factor, ρ. It is defined as the ratio between the experimental value and the predicted one (or the ratio between the frequency factor and the collision frequency), and it is most often less than unity.

$$\rho = \frac{A_{observed}}{Z_{calculated}}$$

Usually, the more complex the reactant molecules, the lower the steric factor. Nevertheless, some reactions exhibit steric factors greater than unity: the harpoon reactions, which involve atoms that exchange electrons, producing ions. The deviation from unity can have different causes: the molecules are not spherical, so different geometries are possible; not all the kinetic energy is delivered into the right spot; the presence of a solvent (when applied to solutions), etc.

Experimental rate constants compared to the ones predicted by collision theory for gas phase reactions			
Reaction	A (Azra frequency factor)	Z (collision frequency)	Steric factor
$2ClNO \rightarrow 2Cl + 2NO$	$9.4 \ 10^9$	$5.9 \ 10^{10}$	0.16
$2ClO \rightarrow Cl_2 + O_2$	$6.3 \ 10^7$	$2.5 \ 10^{10}$	$2.3 \ 10^{-3}$
$H_2 + C_2H_4 \rightarrow C_2H_6$	$1.24 \ 10^6$	$7.3 \ 10^{11}$	$1.7 \ 10^{-6}$
$Br_2 + K \rightarrow KBr + Br$	10^{12}	$2.1 \ 10^{11}$	4.3

Collision theory can be applied to reactions in solution; in that case, the *solvent cage* has an effect on the reactant molecules and several collisions can take place in a single encounter, which leads to predicted preexponential factors being too large. ρ values greater than unity can be attributed to favorable entropic contributions.

Experimental rate constants compared to the ones predicted by collision theory for reactions in solution				
Reaction	Solvent	A 10^{-11}	Z 10^{-11}	Steric factor
$C_2H_5Br + OH^-$	C_2H_5OH	4.30	3.86	1.11
$C_2H_5O^- + CH_3I$	C_2H_5OH	2.42	1.93	1.25
$ClCH_2CO_2^- + OH^-$	water	4.55	2.86	1.59
$C_3H_6Br_2 + I^-$	CH_3OH	1.07	1.39	0.77
$HOCH_2CH_2Cl + OH^-$	water	25.5	2.78	9.17
$4\text{-}CH_3C_6H_4O^- + CH_3I$	ethanol	8.49	1.99	4.27
$CH_3(CH_2)_2Cl + I^-$	$(CH_3)_2CO$	0.085	1.57	0.054
$C_5H_5N + CH_3I$	$C_2H_2Cl_4$	-	-	$2.0 \ 10^{-6}$

Order of Reaction

In chemical kinetics, the order of reaction with respect to a given substance (such as reactant, catalyst or product) is defined as the index, or exponent, to which its concentration term in the rate equation is raised. For the typical rate equation of form $r = k[A]^x[B]^y ...$, where [A], [B], ... are concentrations, the reaction orders (or partial reaction orders) are x for substance A, y for substance B, etc. The *overall* reaction order is the sum x + y + ... For many reactions, the reaction orders are *not* equal to the stoichiometric coefficients.

For example, the chemical reaction between mercury (II) chloride and oxalate ion

$$2HgCl_2 + C_2O_4^{2-} \rightarrow 2Cl^- + 2CO_2 \uparrow + Hg_2Cl_2 \downarrow$$

has the observed rate equation

$$r = k[HgCl_2]^1[C_2O_4^{2-}]^2$$

In this case, the reaction order *with respect to* the reactant $HgCl_2$ is 1 and with respect to oxalate ion is 2; the *overall* reaction order is $1 + 2 = 3$. The reaction orders (here 1 and 2 respectively) differ from the stoichiometric coefficients (2 and 1). Reaction orders can be determined only by experiment. Their knowledge allows conclusions to be drawn about the reaction mechanism, and may help to identify the rate-determining step.

Elementary (single-step) reactions do have reaction orders equal to the stoichiometric coefficients for each reactant. The overall reaction order, i.e. the sum of stoichiometric coefficients of reactants, is always equal to the molecularity of the elementary reaction. Complex (multi-step) reactions may or may not have reaction orders equal to their stoichiometric coefficients.

Orders of reaction for each reactant are often positive integers, but they may also be zero, fractional, or negative.

A reaction can also have an *undefined* reaction order with respect to a reactant if the rate is not simply proportional to some power of the concentration of that reactant; for example, one cannot talk about reaction order in the rate equation for a bimolecular reaction between adsorbed molecules:

$$r = k\frac{K_1K_2C_AC_B}{(1+K_1C_A+K_2C_B)^2}$$

Determination of Reaction Order

Method of Initial Rates

The order of a reaction for each reactant can be estimated from the variation in initial rate with the concentration of that reactant, using the natural logarithm of the typical rate equation

$$\ln r = \ln k + x\ln[A] + y\ln[B] + ...$$

For example, the initial rate can be measured in a series of experiments at different initial concentrations of reactant A with all other concentrations [B], [C], ... kept constant, so that

$$\ln r = x\ln[A] + \text{constant}$$

The slope of a graph of $\ln r$ as a function of $\ln[A]$ then corresponds to the order x with respect to reactant A.

However this method is not always reliable because

1. measurement of the initial rate requires accurate determination of small changes in concentration in short times (compared to the reaction half-life) and is sensitive to errors, and

2. the rate equation will not be completely determined if the rate also depends on substances not present at the beginning of the reaction, such as intermediates or products.

Integral Method

The tentative rate equation determined by the method of initial rates is therefore normally verified by comparing the concentrations measured over a longer time (several half-lives) with the integrated form of the rate equation.

For example, the integrated rate law for a first-order reaction is

$$\ln[A] = -kt + \ln[A]_0,$$

where [A] is the concentration at time t and $[A]_0$ is the initial concentration at zero time. The first-order rate law is confirmed if $\ln[A]$ is in fact a linear function of time. In this case the rate constant k is equal to the slope with sign reversed.

Method of Flooding

The partial order with respect to a given reactant can be evaluated by the method of flooding (or of isolation) of Ostwald. In this method, the concentration of one reactant is measured with all other reactants in large excess so that their concentration remains

essentially constant. For a reaction $a \cdot A + b \cdot B \to c \cdot C$ with rate law: $r = k \cdot [A]^\alpha \cdot [B]^\beta$, the partial order α with respect to A is determined using a large excess of B. In this case

$$r = k' \cdot [A]^\alpha \text{ with } k' = k \cdot [B]^\beta,$$

and α may be determined by the integral method. The order β with respect to B under the same conditions (with B in excess) is determined by a series of similar experiments with a range of initial concentration $[B]_0$ so that the variation of k' can be measured.

First Order

If a reaction rate depends on a single reactant and the value of the exponent is one, then the reaction is said to be first order. In organic chemistry, the class of S_N1 (nucleophilic substitution unimolecular) reactions consists of first-order reactions. For example, in the reaction of aryldiazonium ions with nucleophiles in aqueous solution $ArN_2^+ + X^- \to ArX + N_2$, the rate equation is $r = k[ArN_2^+]$, where Ar indicates an aryl group.

Another class of first-order reactions is radioactive decay processes which are all first order. These are, however, nuclear reactions rather than chemical reactions.

Second Order

A reaction is said to be second order when the overall order is two. The rate of a second-order reaction may be proportional to one concentration squared $r = k[A]^2$, or (more commonly) to the product of two concentrations $r = k[A][B]$. As an example of the first type, the reaction $NO_2 + CO \rightarrow NO + CO_2$ is second-order in the reactant NO_2 and zero order in the reactant CO. The observed rate is given by $r = k[NO_2]^2$, and is independent of the concentration of CO.

The second type includes the class of $S_N 2$ (nucleophilic substitution bimolecular) reactions, such as the alkaline hydrolysis of ethyl acetate:

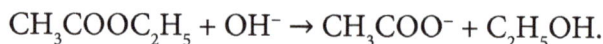

$$CH_3COOC_2H_5 + OH^- \rightarrow CH_3COO^- + C_2H_5OH.$$

This reaction is first-order in each reactant and second-order overall:
$$r = k[CH_3COOC_2H_5][OH^-]$$

If the same hydrolysis reaction is catalyzed by imidazole, the rate equation becomes $r = k[imidazole][CH_3COOC_2H_5]$. The rate is first-order in one reactant (ethyl acetate), and also first-order in imidazole which as a catalyst does not appear in the overall chemical equation.

Pseudo-first Order

If the concentration of a reactant remains constant (because it is a catalyst or it is in great excess with respect to the other reactants), its concentration can be included in the rate constant, obtaining a *pseudo–first-order* (or occasionally pseudo–second-order) rate equation. For a typical second-order reaction with rate equation r = k[A][B], if the concentration of reactant B is constant then r = k[A][B] = k'[A], where the pseudo–first-order rate constant k' = k[B]. The second-order rate equation has been reduced to a pseudo–first-order rate equation, which makes the treatment to obtain an integrated rate equation much easier.

For example, the hydrolysis of sucrose in acid solution is often cited as a first-order reaction with rate r = k[sucrose]. The true rate equation is third-order, r = k[sucrose][H^+][H_2O]; however, the concentrations of both the catalyst H^+ and the solvent H_2O are normally constant, so that the reaction is pseudo–first-order.

Zero Order

For zero-order reactions, the reaction rate is independent of the concentration of a reactant, so that changing its concentration has no effect on the speed of the reaction.

This may occur when there is a bottleneck which limits the number of reactant molecules that can react at the same time, for example if the reaction requires contact with an enzyme or a catalytic surface.

Many enzyme-catalyzed reactions are zero order, provided that the reactant concentration is much greater than the enzyme concentration which controls the rate, so that the enzyme is *saturated*. For example, the biological oxidation of ethanol to acetaldehyde by the enzyme liver alcohol dehydrogenase (LADH) is zero order in ethanol.

Similarly reactions with heterogeneous catalysis can be zero order if the catalytic surface is saturated. For example, the decomposition of phosphine (PH_3) on a hot tungsten surface at high pressure is zero order in phosphine which decomposes at a constant rate.

Fractional Order

In fractional order reactions, the order is a non-integer, which often indicates a chemical chain reaction or other complex reaction mechanism. For example, the pyrolysis of ethanal (CH_3CHO) into methane and carbon monoxide proceeds with an order of 1.5 with respect to ethanal: $r = k[CH_3CHO]^{3/2}$. The decomposition of phosgene ($COCl_2$) to carbon monoxide and chlorine has order 1 with respect to phosgene itself and order 0.5 with respect to chlorine: $r = k[COCl_2][Cl_2]^{1/2}$.

The order of a chain reaction can be rationalized using the steady state approximation for the concentration of reactive intermediates such as free radicals. For the pyrolysis of ethanal, the Rice-Herzfeld mechanism is

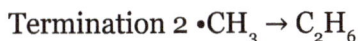

Initiation $CH_3CHO \rightarrow \cdot CH_3 + \cdot CHO$

Propagation $\cdot CH_3 + CH_3CHO \rightarrow CH_3CO\cdot + CH_4$

$CH_3CO\cdot \rightarrow \cdot CH_3 + CO$

Termination $2 \cdot CH_3 \rightarrow C_2H_6$

where $\cdot$ denotes a free radical. To simplify the theory, the reactions of the $\cdot CHO$ to form a second $\cdot CH_3$ are ignored.

In the steady state, the rates of formation and destruction of methyl radicals are equal, so that

$$\frac{d[.CH_3]}{dt} = k_i[CH_3CHO] - k_t[.CH_3]^2 = 0,$$

so that the concentration of methyl radical satisfies

$$[.CH3] \propto [CH_3CHO]^{1/2}$$

The reaction rate equals the rate of the propagation steps which form the main reaction products CH_4 and CO:

$$v = \frac{d[CH_4]}{dt} = k_p[.CH_3][CH_3CHO] \quad \propto \quad [CH_3CHO]^{\frac{3}{2}}$$

in agreement with the experimental order of 3/2.

Mixed Order

More complex rate laws have been described as being *mixed order* if they approximate to the laws for more than one order at different concentrations of the chemical species involved. For example, a rate law of the form $r = k_1[A] + k_2[A]^2$ represents concurrent first order and second order reactions (or more often concurrent pseudo-first order and second order) reactions, and can be described as mixed first and second order. For sufficiently large values of [A] such a reaction will approximate second order kinetics, but for smaller [A] the kinetics will approximate first order (or pseudo-first order). As the reaction progresses, the reaction can change from second order to first order as reactant is consumed.

Another type of mixed-order rate law has a denominator of two or more terms, often because the identity of the rate-determining step depends on the values of the concentrations. An example is the oxidation of an alcohol to a ketone by hexacyanoferrate (III) ion $[Fe(CN)_6^{3-}]$ with ruthenate (VI) ion (RuO_4^{2-}) as catalyst. For this reaction, the rate of disappearance of

hexacyanoferrate (III) is $r = \dfrac{[Fe(CN)_6]^{2-}}{k_\alpha + k_\beta[Fe(CN)_6]^{2-}}$

This is zero-order with respect to hexacyanoferrate (III) at the onset of the reaction (when its concentration is high and the ruthenium catalyst is quickly regenerated), but changes to first-order when its concentration decreases and the regeneration of catalyst becomes rate-determining.

Notable mechanisms with mixed-order rate laws with two-term denominators include:

- Michaelis-Menten kinetics for enzyme-catalysis: first-order in substrate (second-order overall) at low substrate concentrations, zero order in substrate (first-order overall) at higher substrate concentrations; and

- the Lindemann mechanism for unimolecular reactions: second-order at low pressures, first-order at high pressures.

Negative Order

A reaction rate can have a negative partial order with respect to a substance. For example, the conversion of ozone (O_3) to oxygen follows the rate equation $r = k \dfrac{[O_3]^2}{[O_2]}$ in an excess of oxygen. This corresponds to second order in ozone and order (-1) with respect to oxygen.

When a partial order is negative, the overall order is usually considered as undefined. In the above example for instance, the reaction is not described as first order even though the sum of the partial orders is 2 + (-1) = 1, because the rate equation is more complex than that of a simple first-order reaction.

Chain Reaction

A chain reaction is a sequence of reactions where a reactive product or by-product causes additional reactions to take place. In a chain reaction, positive feedback leads to a self-amplifying chain of events.

Chain reactions are one way in which systems which are in thermodynamic non-equilibrium can release energy or increase entropy in order to reach a state of higher entropy. For example, a system may not be able to reach a lower energy state by releasing energy into the environment, because it is hindered or prevented in some way from taking the path that will result in the energy release. If a reaction results in a small energy release making way for more energy releases in an expanding chain, then the system will typically collapse explosively until much or all of the stored energy has been released.

A macroscopic metaphor for chain reactions is thus a snowball causing a larger snowball until finally an avalanche results ("snowball effect"). This is a result of stored gravitational potential energy seeking a path of release over friction. Chemically, the equivalent to a snow avalanche is a spark causing a forest fire. In nuclear physics, a single stray neutron can result in a prompt critical event, which may finally be energetic enough for a nuclear reactor meltdown or (in a bomb) a nuclear explosion.

Chemical Chain Reactions

History

In 1913 the German chemist Max Bodenstein first put forth the idea of chemical chain reactions. If two molecules react, not only molecules of the final reaction products are formed, but also some unstable molecules which can further react with the parent molecules with a far larger probability than the initial reactants. In the new reaction, further unstable molecules are formed besides the stable products, and so on.

In 1918, Walther Nernst proposed that the photochemical reaction of hydrogen and chlorine is a chain reaction in order to explain the large quantum yield, meaning that one photon of light is responsible for the formation of as many as 10^6 molecules of the product HCl. He suggested that the photon dissociates a Cl_2 molecule into two Cl atoms which each initiate a long chain of reaction steps forming HCl.

In 1923, Danish and Dutch scientists Christian Christiansen and Hendrik Anthony Kramers, in an analysis of the formation of polymers, pointed out that such a chain reaction need not start with a molecule excited by light, but could also start with two molecules colliding violently due to thermal energy as previously proposed for initiation of chemical reactions by van' t Hoff.

Christiansen and Kramers also noted that if, in one link of the reaction chain, two or more unstable molecules are produced, the reaction chain would branch and grow. The result is in fact an exponential growth, thus giving rise to explosive increases in reaction rates, and indeed to chemical explosions themselves. This was the first proposal for the mechanism of chemical explosions.

A quantitative chain chemical reaction theory was created by Soviet physicist Nikolay Semyonov in 1934. Semyonov shared the Nobel Prize in 1956 with Sir Cyril Norman Hinshelwood, who independently developed many of the same quantitative concepts.

Typical Steps

The main types of steps in chain reaction are of the following types.

- Initiation (formation of active particles or chain carriers, often free radicals, in either a thermal or a photochemical step)

- Propagation (may comprise several elementary steps in a cycle, where the active particle through reaction forms another active particle which continues the reaction chain by entering the next elementary step). In effect the active particle serves as a catalyst for the overall reaction of the propagation cycle. Particular cases are:

  1. chain branching (a propagation step which forms more new active particles than enter the step);

  2. chain transfer (a propagation step in which the active particle is a growing polymer chain which reacts to form an inactive polymer whose growth is terminated and an active small particle (such as a radical), which may then react to form a new polymer chain).

- Termination (elementary step in which the active particle loses its activity; e. g. by recombination of two free radicals).

The *chain length* is defined as the average number of times the propagation cycle is repeated, and equals the overall reaction rate divided by the initiation rate.

Some chain reactions have complex rate equations with fractional order or mixed order kinetics.

Detailed Example: the Hydrogen-bromine Reaction

The reaction $H_2 + Br_2 \rightarrow 2\,HBr$ proceeds by the following mechanism:

- Initiation

  $Br_2 \rightarrow 2\,Br\bullet$ (thermal) or $Br_2 + h\nu \rightarrow 2\,Br\bullet$ (photochemical)

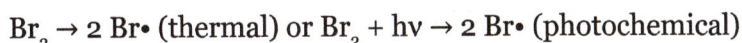

  each Br atom is a free radical, indicated by the symbol « • » representing an unpaired electron.

- Propagation (here a cycle of two steps)

  $Br\bullet + H_2 \rightarrow HBr + H\bullet$

  $H\bullet + Br_2 \rightarrow HBr + Br\bullet$

  the sum of these two steps corresponds to the overall reaction $H_2 + Br_2 \rightarrow 2\,HBr$, with catalysis by Br• which participates in the first step and is regenerated in the second step.

- Retardation (inhibition)

  $H\bullet + HBr \rightarrow H_2 + Br\bullet$

  this step is specific to this example, and corresponds to the first propagation step in reverse.

- Termination $2\,Br\bullet \rightarrow Br_2$

  recombination of two radicals, corresponding in this example to initiation in reverse.

As can be explained using the steady-state approximation, the thermal reaction has an initial rate of fractional order (3/2), and a complete rate equation with a two-term denominator (mixed-order kinetics).

Further Chemical Examples

- The reaction $2\,H_2 + O_2 \rightarrow 2\,H_2O$ provides an example of chain branching. The propagation is a sequence of two steps whose net effect is to replace an H atom by another H atom plus two OH radicals. This leads to an explosion under certain conditions of temperature and pressure.

  o $H + O_2 \rightarrow OH + O$

  o $O + H_2 \rightarrow OH + H$

- In chain-growth polymerization, the propagation step corresponds to the elongation of the growing polymer chain. Chain transfer corresponds to transfer of the activity from this growing chain, whose growth is terminated, to another molecule which may be a second growing polymer chain. For polymerization, the kinetic chain length defined above may differ from the degree of polymerization of the product macromolecule.

- Polymerase chain reaction, a technique used in molecular biology to amplify (make many copies of) a piece of DNA by *in vitro* enzymatic replication using a DNA polymerase.

Nuclear Chain Reactions

A *nuclear* chain reaction was proposed by Leo Szilard in 1933, shortly after the neutron was discovered, yet more than five years before nuclear fission was first discovered. Szilárd knew of *chemical* chain reactions, and he had been reading about an energy-producing nuclear reaction involving high-energy protons bombarding lithium, demonstrated by John Cockcroft and Ernest Walton, in 1932. Now, Szilárd proposed to use neutrons theoretically produced from certain nuclear reactions in lighter isotopes, to induce further reactions in light isotopes that produced more neutrons. This would in theory produce a chain reaction at the level of the nucleus. He did not envision fission as one of these neutron-producing reactions, since this reaction was not known at the time. Experiments he proposed using beryllium and indium failed.

Later, after fission was discovered in 1938, Szilárd immediately realized the possibility of using neutron-induced fission as the particular nuclear reaction necessary to create a chain-reaction, so long as fission also produced neutrons. In 1939, with Enrico Fermi, Szilárd proved this neutron-multiplying reaction in uranium. In this reaction, a neutron plus a fissionable atom causes a fission resulting in a larger number of neutrons than the single one that was consumed in the initial reaction. Thus was born the practical nuclear chain reaction by the mechanism of neutron-induced nuclear fission.

Specifically, if one or more of the produced neutrons themselves interact with other fissionable nuclei, and these also undergo fission, then there is a possibility that the macroscopic overall fission reaction will not stop, but continue throughout the reaction material. This is then a self-propagating and thus self-sustaining chain reaction. This is the principle for nuclear reactors and atomic bombs.

Demonstration of a self-sustaining nuclear chain reaction was accomplished by Enrico Fermi and others, in the successful operation of Chicago Pile-1, the first artificial nuclear reactor, in late 1942.

Electron Avalanche in Gases

An electron avalanche happens between two unconnected electrodes in a gas when an electric field exceeds a certain threshold. Random thermal collisions of gas atoms may result in a few free electrons and positively charged gas ions, in a process called impact ionization. Acceleration of these free electrons in a strong electric field causes them to gain energy, and when they impact other atoms, the energy causes release of new free electrons and ions (ionization), which fuels the same process. If this process happens faster than it is naturally quenched by ions recombining, the new ions multiply in successive cycles until the gas breaks down into a plasma and current flows freely in a discharge.

Electron avalanches are essential to the dielectric breakdown process within gases. The process can culminate in corona discharges, streamers, leaders, or in a spark or continuous electric arc that completely bridges the gap. The process may extends to huge sparks — streamers in lightning discharges propagate by formation of electron avalanches created in the high potential gradient ahead of the streamers' advancing tips. Once begun, avalanches are often intensified by the creation of photoelectrons as a result of ultraviolet radiation emitted by the excited medium's atoms in the aft-tip region. The extremely high temperature of the resulting plasma cracks the surrounding gas molecules and the free ions recombine to create new chemical compounds.

The process can also be used to detect radiation that initiates the process, as the passage of a single particles can amplified to large discharges. This is the mechanism of a Geiger counter and also the visualization possible with a spark chamber and other wire chambers.

Avalanche Breakdown in Semiconductors

An avalanche breakdown process can happen in semiconductors, which in some ways conduct electricity analogously to a mildly ionized gas. Semiconductors rely on free electrons knocked out of the crystal by thermal vibration for conduction. Thus, unlike metals, semiconductors become better conductors the higher the temperature. This sets up conditions for the same type of positive feedback—heat from current flow causes temperature to rise, which increases charge carriers, lowering resistance, and causing more current to flow. This can continue to the point of complete breakdown of normal resistance at a semiconductor junction, and failure of the device (this may be temporary or permanent depending on whether there is physical damage to the crystal). Certain devices, such as avalanche diodes, deliberately make use of the effect.

Chain Branching Explosion

Explosion

Very rapid combustion of fuel and oxidizer, leading to violent release of energy.

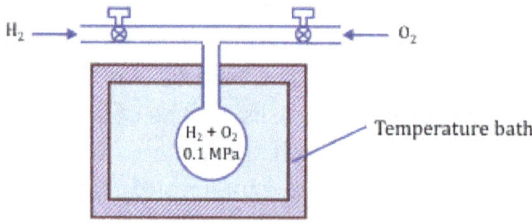

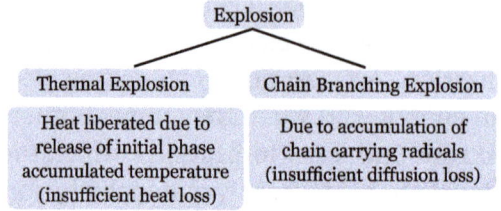

- To begin with, stoichiometric mixture of H_2 and O_2 is kept in a container.

- Temperature is increased beyond 773 K.

- Result: Very rapid chemical reaction with explosion.

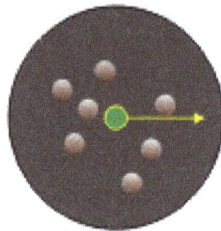

Low pressure
(No exlosion)

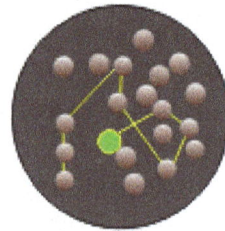

High pressure
(Exlosion)

Regimes in the explosion chart

1. First limit

2. Second limit

3. Third limit

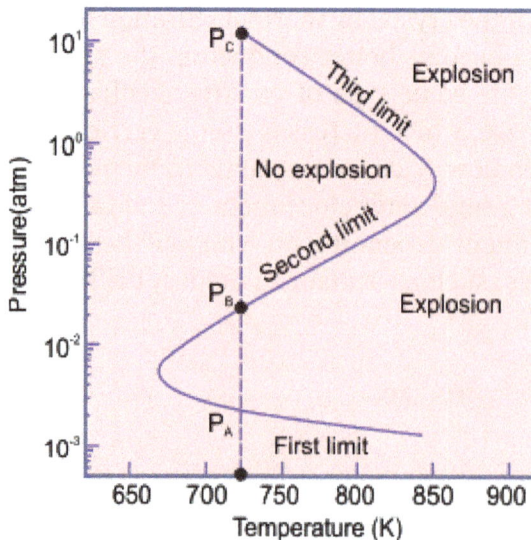

Multistep Reaction Mechanism

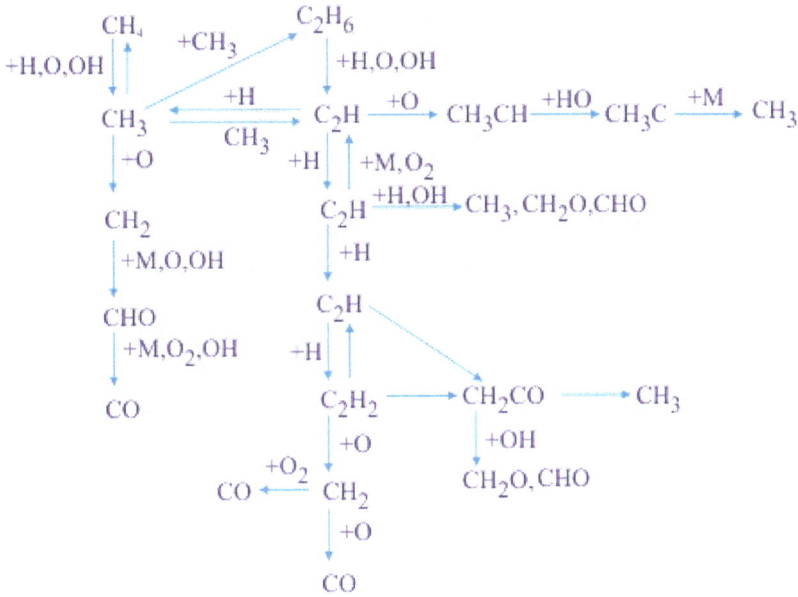

Quasi-Steady State Approximation

- Radicals are formed during combustion
- Half life period of radicals - Very small
- Rate of formation = Rate of destruction

Relate radical concentration with

measurable concentration of other species

Consider the two step chain reaction,

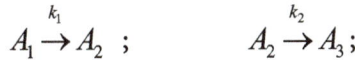

$$A_1 \xrightarrow{k_1} A_2 \; ; \qquad A_2 \xrightarrow{k_2} A_3 ;$$

Reaction rate of the three species,

$$\frac{dC_{A_1}}{dt} = -k_1 C_{A_1} \qquad \frac{dC_{A_2}}{dt} = k_1 C_{A_1} - k_2 C_{A_2} \qquad \frac{dC_{A_3}}{dt} = k_2 C_{A_2}$$

Initial condition, $at\ t = 0; C_{A_1} = C_{A_{1,in}}; \; C_{A_2} = 0; \; C_{A_3} = 0$

Applying initial condition,

$$C_{A_1} = C_{A_{1,in}} e^{(-k_2 t)}$$

$$C_{A_1} = C_{A_1,in} \frac{k_1}{k_1 - k_2} [e^{(-k_2 t)} - e^{(-k_1 t)}]$$

$$C_{A_3} = C_{A_1,0} \left[1 - \frac{k_1}{k_1 - k_2} e^{(-k_2 t)} + \frac{k_2}{k_1 - k_2} e^{(-k_1 t)} \right]$$

Applying QSSA method to A2,

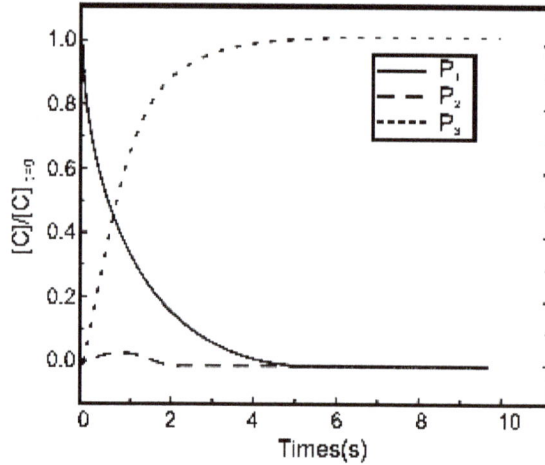

$$\frac{dC_{A_2}}{dt} = k_1 C_{A_1} - k_2 C_{A_2} \approx 0$$

$$k_1 C_{A_1} = k_2 C_{A_2} ; \qquad\qquad \frac{dC_{A_3}}{dt} = k_1 C_{A_1,in} e^{(-k_1 t)} ;$$

$$\frac{dC_{A_3}}{dt} = k_1 C_{A_1} ; \qquad\qquad C_{A_3} = C_{A_1,in} [1 - e^{(-k_1 t)}] ;$$

QSSA method predicts the concentration of species especially when $k_2 \gg k_1$

Partial Equilibrium Approximation (PEA)

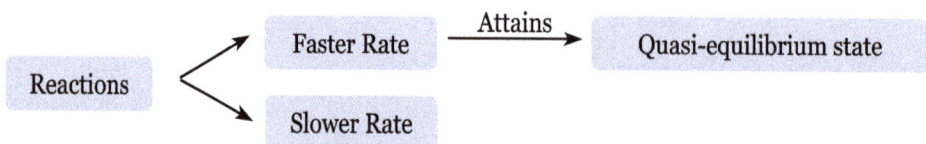

PEA expresses concentration of unknown species in terms of known concentrations.

Consider NO formation mechanism,

$$O + N_2 \Leftrightarrow NO + N \qquad\qquad k_f = 2 \times 10^{14} e^{(-315/RT)}$$

Reaction Rate (RR) for NO species:

$$\frac{dC_{NO}}{dt} = k_f C_O C_{N_2}$$

Note 1:   O and N_2 concentration are required to determine RR

Note 2:   Rate of formation and destruction of O is very high

Difficult to measure the concentration of O:

Step 1:

Assume partial equilibrium for O_2 molecule

PEA	QSSA
Reaction attains steady state	Species attains steady state

Step 2:

Relate O_2 molecule to O by

$$O_2 \Leftrightarrow 2O$$

Equilibrium constant

$$K_c = \frac{C_O^2}{C_{O_2}}$$

Note

- Thermal NO – Less dependent on CO_2

- Thermal NO – Temperature dependent

- NO reduction – By reducing combustion temperature

Reaction Rate (RR) for NO species:

$$\frac{dC_{NO}}{dt} = k_f (K_c C_{o_2})^{0.5} C_{N_2}$$

> **Caution**
>
> - In real situation, particular reaction may not attain equilibrium.
> - PEA provides satisfactory results only at high temperature.

Partial Equilibrium

Partial equilibrium is a condition of economic equilibrium which takes into consideration only a part of the market, ceteris paribus, to attain equilibrium.

As defined by George Stigler, "A partial equilibrium is one which is based on only a restricted range of data, a standard example is price of a single product, the prices of all other products being held fixed during the analysis."

The supply and demand model is a partial equilibrium model where the clearance on the market of some specific goods is obtained independently from prices and quantities in other markets. In other words, the prices of all substitutes and complements, as well as income levels of consumers, are taken as given. This makes analysis much simpler than in a general equilibrium model which includes an entire economy.

Here the dynamic process is that prices adjust until supply equals demand. It is a powerfully simple technique that allows one to study equilibrium, efficiency and comparative statics. The stringency of the simplifying assumptions inherent in this approach make the model considerably more tractable, but may produce results which, while seemingly precise, do not effectively model real-world economic phenomena.

Partial equilibrium analysis examines the effects of policy action in creating equilibrium only in that particular sector or market which is directly affected, ignoring its effect in any other market or industry assuming that they being small will have little impact if any.

Hence this analysis is considered to be useful in constricted markets.

Léon Walras first formalized the idea of a one-period economic equilibrium of the general economic system, but it was French economist Antoine Augustin Cournot and English political economist Alfred Marshall who developed tractable models to analyze an economic system.

Assumptions

1. Commodity price is given and constant for the consumers.

2. Consumers' taste and preferences, habits, incomes are also considered to be constant.

3. Prices of prolific resources of a commodity and that of other related goods (substitute or complementary) are known as well as constant.

4. Industry is easily availed with factors of production at a known and constant price compliant with the methods of production in use.

5. Prices of the products that the factor of production helps in producing and the price and quantity of other factors are known and constant.

6. There is perfect mobility of factors of production between occupation and places.

The above-mentioned points relate to a perfectly competitive market but can be further extended to monopolistic competition, oligopoly, monopoly and monopsony markets.

Applications

Applications of partial equilibrium discusses, when does an individual, a firm, an industry, factors of production attain their equilibrium points:

1. A consumer is in a state of equilibrium when they achieve maximum aggregate satisfaction on the expenditure that they make depending on the set of conditions relating to his tastes and preferences, income, price and supply of the commodity etc.

2. Producers' equilibrium occurs when they maximize their net profit subject to a given set of economic situations.

3. A firm's equilibrium point is when it has no inclination in changing its production.

   o In the short run: Marginal Revenue = Marginal Cost.

Algebraically MR=MC

   o In long run: Long run Marginal Cost = Marginal Revenue = Average Revenue = Long run Average Cost

Algebraically LMC=MR=AR=LAC at its minimum are the conditions of equilibrium. It means that a firm is earning only a "normal profit" and has no intension to leave the industry.

1. Equilibrium for an industry happens when there is normal profit made by an industry It is such a situation when no new firm wants to enter into it and the existing firm does not want to exit.

Only one price prevails in the market for a single product where the quantity of goods purchased by a buyer = total quantity produced by different firms. All the firms produces till that level where Marginal Cost=Marginal Revenue, and sells the product at market price ruling at that point of time.

1. Factors of production, i.e., land, labor, capital, and entrepreneurs are in equilibrium when they are paid the maximum possible so as maximize the income. Here the Price = Marginal Revenue Product.

At this price it does not have any enticement to look for employment anywhere else.

The quantity of factors which its owners want to sell should be equal to the quantity which the entrepreneurs are ready to hire.

Limitations

1. It is restricted to one particular portion of the economy.

2. It lacks the ability to study the interrelations of all the parts of the economy.

3. This analysis will fail if the improbable assumptions, which disconnect the study of specific market from the rest of the economy, are not taken into consideration.

4. It has been unsuccessful in explaining the outcome of economic disturbance in the market that leads to demand and supply changes, moving from one market to another and thus instigating second- and third-order waves of change in the whole economy.

Welfare Effect of Trade Policies

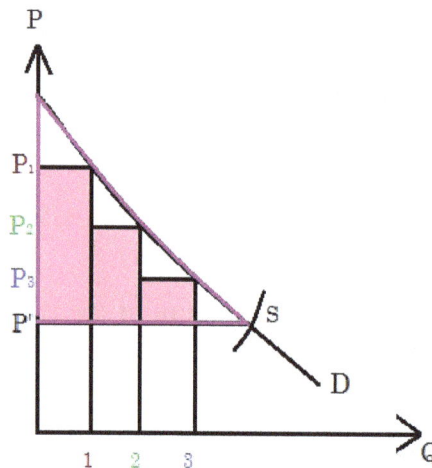

Graph showing Consumer Surplus in the market, P refers to Price on the Y-axis, Q refers to Quantity on X-axis, D- Demand Curve, S-Supply Curve.

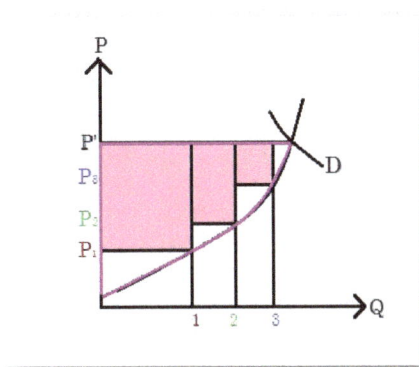

Graph showing Producer Surplus in the market, P refers to Price on the Y-axis, Q refers to Quantity on X-axis, D- Demand Curve, S-Supply Curve.

In partial equilibrium the welfare effects on consumers who purchase and the producers who produce in the market is distinguished by consumer surplus and producer surplus.

Consumer Surplus

The amount that a consumer is ready to pay for a particular good minus the amount that the consumer actually pays. The amount that the consumer is willing to pay has to be greater.

In the graph given here, P_1 is the price that a consumer is ready to pay for a particular product. But the producer may reduce the price to P_2 expecting that either more people would buy at the reduced rate, or the person who was ready to pay P_1 will purchase more of the same. The producer may further reduce the price to P_3, again expecting more buyers or the same buyers purchasing more.

The price keeps on falling until P', where the demand and the supply curves intersect: their intersection is the equilibrium point. Hence the consumer surplus for first consumer can be calculated as P_1 - P', decreasing for the second consumer to P_2 - P', and so on. Thus the total consumer surplus in the market can be obtained by summing up the three rectangles. The triangle with the purple outline to the left indicates that area.

Producer Surplus

Amount that a producer finally receives by selling a particular product minus the amount the producer is ready to accept for that good. The amount that the producer receives should be greater.

If only one unit of the commodity was demanded at the price P_1, this becomes the price which the producer expects to receive. But if two units are demanded, the minimum price at which the producer would be ready to increase the supply shifts to P_2. This continues and the final price that ultimately prevails in the market is P', the price which is obtained by the intersection of the demand and supply curve in the market. The producer's surplus here would be initial price minus the final price. And total consumer surplus in the market will be summation of the three rectangles.

Difference between Partial and General Equilibrium

Partial Equilibrium	General Equilibrium
• Developed by Alfred Marshall.	• Léon Walras was first to develop it.
• Related to single variable	• More than one variable or economy as a whole is taken into consideration
• Based on two assumptions: 1. Ceteris Paribus 2. Other sectors are not affected due to change in one sector.	• It is based on the assumption that various sectors are mutually interdependent. There is an effect on other sectors due to change in one.
• Other things remaining constant, price of a good is determined	• Prices of goods are determined simultaneously and mutually. Hence all product and factor markets are simultaneously in equilibrium.

Global Kinetics

Single step methane combustion:

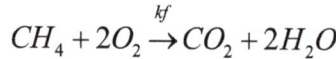

$$CH_4 + 2O_2 \xrightarrow{kf} CO_2 + 2H_2O$$

Overall reaction rate (CH4)

$$\frac{dC_{CH_4}}{dt} = A_e{}^{(-E/R_uT)} C_{CH_4}^m C_{O_2}^n$$

Global kinetic scheme for an arbitrary hydrocarbon (CxHy):

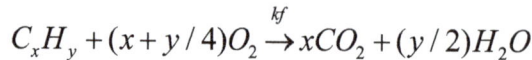

$$C_xH_y + (x + y/4)O_2 \xrightarrow{kf} xCO_2 + (y/2)H_2O$$

Overall reaction rate (C_xH_y)

$$\frac{dC_xH_y}{dt} = -A_f e^{(-E/R_uT)} C_{C_xh_y}^m C_{O_2}^m$$

References

- McGill and King (1995). Engineering Mechanics, An Introduction to Dynamics (3rd ed.). PWS Publishing Company. ISBN 0-534-93399-8

- Z.Y.Wang (2016). "Generalized momentum equation of quantum mechanics". Optical and Quantum Electronics. 48 (2): 1–9. doi:10.1007/s11082-015-0261-8

- Hand, Louis N.; Finch, Janet D. (1998). Analytical mechanics (7th print ed.). Cambridge, England: Cambridge University Press. Chapter 4. ISBN 9780521575720

- "Lab Note #106 Environmental Impact of Arc Suppression". Arc Suppression Technologies. April 2011. Retrieved March 15, 2012

- Barnett, Stephen M. (2010). "Resolution of the Abraham-Minkowski Dilemma". Physical Review Letters. 104 (7). Bibcode:2010PhRvL.104g0401B. PMID 20366861. doi:10.1103/PhysRevLett.104.070401

- Gubbins, David (1992). Seismology and plate tectonics (Repr. (with corr.) ed.). Cambridge [England]: Cambridge University Press. p. 59. ISBN 0521379954

- Wang Zhong-Yue; Wang Pin-Yu; Xu Yan-Rong (2011). "Crucial experiment to resolve Abraham-Minkowski Controversy". Optik. 122 (22): 1994–1996. Bibcode:2011Optik.122.1994W. doi:10.1016/j.ijleo.2010.12.018

- Scott, J.F. (1981). The Mathematical Work of John Wallis, D.D., F.R.S. Chelsea Publishing Company. p. 111. ISBN 0-8284-0314-7

- McIntyre, M. E. (1981). "On the 'wave momentum' myth". J. Fluid. Mech. 106: 331–347. Bibcode:1981JFM...106..331M. doi:10.1017/s0022112081001626

- I. Tinoco, K. Sauer and J.C. Wang Physical Chemistry. Principles and Applications in Biological Sciences. (3rd ed., Prentice-Hall 1995) p.328-9 ISBN 0-13-186545-5

- Espinoza, Fernando (2005). "An analysis of the historical development of ideas about motion and its implications for teaching". Physics Education. 40 (2): 141. Bibcode:2005PhyEd..40..139E. doi:10.1088/0031-9120/40/2/002

5

Premixed Flame in Combustion

A pre-mixed flame is formed during combustion of mixture of fuel and an oxidizer. It is necessary that the elements are mixed before they are made to combust. A pre-mixed flame can be characterized by its burning velocity. The topics discussed in the chapter are of great importance to broaden the existing knowledge on combustion.

Premixed Flame

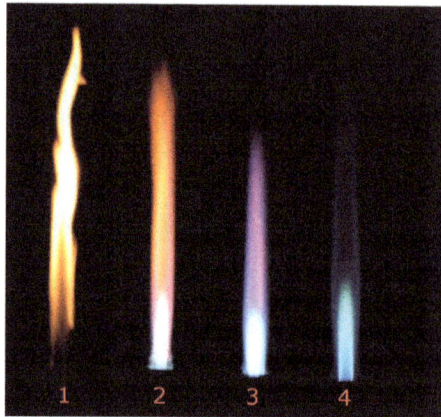

Different flame types of a Bunsen burner depend on oxygen supply. On the left a rich fuel mixture with no premixed oxygen produces a yellow sooty diffusion flame, and on the right a lean fully oxygen premixed flame produces no soot and the flame color is produced by molecular radical band emission.

A premixed flame is a flame formed under certain conditions during the combustion of a premixed charge (also called pre-mixture) of fuel and oxidiser. Since the fuel and oxidiser — the key chemical reactants of combustion — are available throughout a homogeneousstoichiometric premixed charge, the combustion process once initiated sustains itself by way of its own heat release. The majority of the chemical transformation in such a combustion process occurs primarily in a thin interfacial region which separates the unburned and the burned gases. The premixed flame interface propagates through the mixture until the entire charge is depleted. The propagation speed of a premixed flame is known as the flame speed (or burning velocity) which depends on the convection-diffusion-reaction balance within the flame, i.e. on its inner chemical structure. The premixed flame is characterised as laminar or turbulent depending on the velocity distribution in the unburned pre-mixture (which provides the medium of propagation for the flame).

Premixed Flame Propagation

Laminar

Under controlled conditions (typically in a laboratory) a laminar flame may be formed in one of several possible flame configurations. The inner structure of a laminar premixed flame is composed of layers over which the decomposition, reaction and complete oxidation of fuel occurs. These chemical processes are much faster than the physical processes such as vortex motion in the flow and, hence, the inner structure of a laminar flame remains intact in most circumstances. The constitutive layers of the inner structure correspond to specified intervals over which the temperature increases from the specified unburned mixture up to as high as the adiabatic flame temperature (AFT). In the presence of volumetric heat transfer and/or aerodynamic stretch, or under the development intrinsic flame instabilities, the extent of reaction and, hence, the temperature attained across the flame may be different from the AFT.

Laminar Burning Velocity

Variations in local propagation speed of a laminar flame arise due to what is called flame stretch. Flame stretch may be due to aerodynamic strain or flame surface curvature; the difference in the propagation speed from the corresponding laminar speed is a function of these effects and may be written as

$$\frac{u_L - u_F}{u_L} = Ka_s Ma_s + Ka_c Ma_c$$

where $Ka_s Ma_s$ is the strain contribution and the term $Ka_c Ma_c$ is the curvature contribution to the flame stretch. Stretch affects arise due to convective-diffusive-reactive imbalances within the inner flame structure and may vary depending on the fuel and equivalence ratio.

Turbulent

In practical scenarios, turbulence is inevitable and, under moderate conditions, turbulence aids the premixed burning process as it enhances the mixing process of fuel and oxidiser. If the premixed charge of gases is not homogeneously mixed, the variations on equivalence ratio may affect the propagation speed of the flame. In some cases, this is desirable as in stratified combustion of blended fuels.

A turbulent premixed flame can be assumed to propagate as a surface composed of an ensemble of laminar flames so long as the processes that determine the inner structure of the flame are not affected. Under such conditions, the flame surface is wrinkled by virtue of turbulent motion in the premixed gases increasing the surface area of the flame. The wrinkling process increases the burning velocity of the turbulent premixed flame in comparison to its laminar counter-part.

The propagation of such a premixed flame may be analysed using the field equation for a scalar G as

$$\frac{\partial G}{\partial t} + \mathbf{v} \cdot \nabla G = u_f \, |\nabla G|,$$

which is defined such that the level-sets of G represent the various interfaces within the premixed flame propagating with a uniform velocity u_F. This, however, is typically not the case as the propagation speed of the interface (with resect to unburned mixture) varies from point to point due to the aerodynamic stretch induced due to gradients in the velocity field.

Under contrasting conditions, however, the inner structure of the premixed flame may be entirely disrupted causing the flame to extinguish either locally (known as local extinction) or globally (known as global extinction or blow-off). Such opposing cases govern the operation of practical combustion devices such as SI engines as well as aero-engine afterburners. The prediction of the extent to which the inner structure of flame is affected in turbulent flow is a topic of extensive research.

Premixed Flame Configuration

Bunsen Flame

In a Bunsen flame, a steady flow rate is provided which matches the flame speed so as to stabilize the flame. If the flow rate is below the flame speed, the flame will move upstream until the fuel is consumed or until it encounters a flame holder. If the flow rate is equal to the flame speed, we would expect a stationary flat flame front normal to the flow direction. If the flow rate is above the flame speed, the flame front will become conical such that the component of the velocity vector normal to the flame front is equal to the flame speed.

Stagnation Flame

Here, the pre-mixed gases flow in such a way so as to form a region of stagnation (zero velocity) where the flame may be stabilized.

Spherical Flame

In this configuration, the flame is typically initiated by way of a spark within a homogeneous pre-mixture. The subsequent propagation of the developed premixed flame occurs as a spherical front until the mixture is transformed entirely or the walls of the combustion vessel are reached.

Applications

Since the equivalence ratio of the premixed gases may be controlled, premixed combustion offers a means to attain low temperatures and, thereby, reduce NOx emissions. Due to improved mixing in comparison with diffusion flames, soot formation is mitigated as well. Premixed combustion has therefore gained significance in recent times. The uses involve lean-premixed-prevaporized (LPP) gas turbines and SI engines.

One-dimensional Combustion Wave

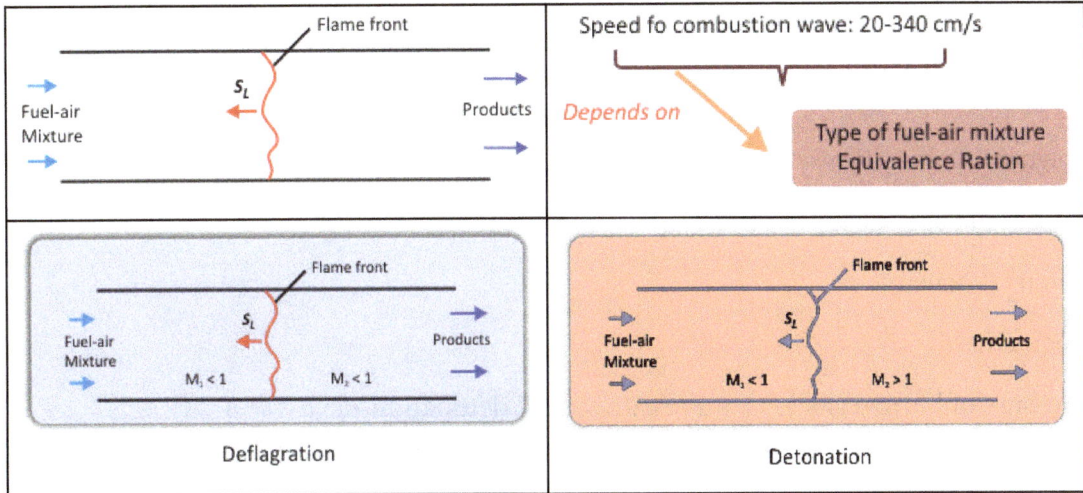

Deflagration

Detonation

Analysis of 1D Flame

Continuity Equation : $\rho_1 V_1 = \rho_1 V_1 = \dot{m}$

State Equations: $P_1 = \rho_1 R T_1$

$$P_2 = \rho_2 R T_2$$

Momentum Equation : $\rho_1 + \rho_1 V_1^2 = P_2 + \rho_1 V_1^2$

Energy Equation : $C_{p1} T_1 + \dfrac{V_1^2}{2} + q = C_{p2} T_2 + \dfrac{V_2^2}{2}$

ρ, V, P, T are the density, velocity, pressure and temperature

q is the heat release per unit mass $= \sum Y_i \Delta h_{f,i}^0$

Y_i is the mass fraction of ith species

$\Delta h_{f,i}^0$ heat of formation of ith species

Combining Continuity and Momentum Equations and expressing them in terms of Mach number,

$$P_1^2 V_1^2 = \frac{P_2 - P_1}{\left(\dfrac{1}{\rho_1} - \dfrac{1}{\rho_2}\right)} = \dot{m}^2 \qquad \longrightarrow \qquad \text{Rayleigh Relation} \quad M_1^2 = \frac{1}{\gamma}\left(\frac{P_1/P_2 - 1}{1 - \rho_1/\rho_2}\right)$$

Rearranging the energy equation, we can get

$$q = \frac{\gamma}{\gamma - 1}\left(\frac{P_2}{\rho_2} - \frac{P_1}{\rho_1}\right) - \frac{1}{2}(P_2 - P_1)\left(\frac{1}{\rho_1} + \frac{1}{\rho_2}\right) \qquad \longrightarrow \quad \text{Rankine - Hugoniot Relation}$$

Hugoniot Curve

HugoniotCurve : P_2 vs. $\dfrac{1}{\rho_2}$ for a

fixed value of q, inlet pressure D_1,

and inlet density ρ_1

Region I:

Pressure of burned gas (P_2) > Pressure of C-J detonation wave (P_U);

 Strong detonation

$\rightarrow$ Gasvelocity relative to wave front is slowed

to subsonic speed

$$\rightarrow M_2 < 1.0$$

$\rightarrow$ Pressure and density increases significantly

for $P_2 \rightarrow \infty, M_1$ will be ∞; rarely observed

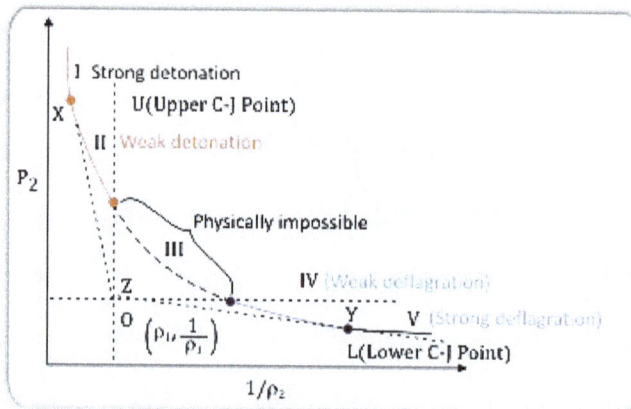

O: Origin of the plot

Region II:

Pressure of burned gas Pressure of $(P_2) <$

C-J detonation wave (P_U); Weak detonation

→ Gas velocity relative to wave front is slowed to subsonic speed

→ Burned gas velocity > speed of sound at isochoric condition $\left(\dfrac{1}{\rho_2} \approx \dfrac{1}{\rho_1} \right)$ weak detonation attains infinite velocity

Region III:

In this region $P_2 < P_1$; Therfore $\dfrac{1}{\rho_2} \gg \dfrac{1}{\rho_1}$

Hence M_1 in this region is imaginary and physically impossible

Structure of 1D Premixed Flame

Preheat zone: Negligibly small heat release

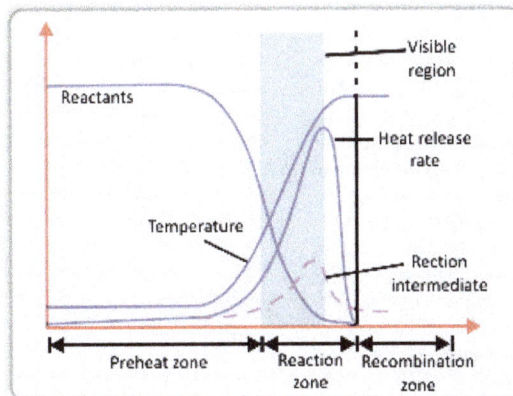

- Certain chemical reactions take place in this zone.

Reaction zone:

- most of the chemical energy is released in the form of heat.

- Decomposition of fuel takes place, leading to intermediate radical formation.

- Reaction zone is very thin as compared to the preheat zone.

- Temperature gradient and concentration gradient are high.

Recombination zone:

- CO_2 and H_2O are formed

- No heat release in this zone

Laminar Flame Theory

Assumptions:

- 1D, steady, inviscid flow.

- Flame is quite thin.

- Ignition temperature is very close to flame temperature.

- No heat loss including radiation; $\Rightarrow$ Adiabatic flame.

- Pressure difference across the flame is negligibly small.

- Binary diffusion, Fourier and Fick's law are valid.

- Unity Lewis number.

- Constant transport properties $\left(k_g, C_P, \mu, D \sim constant \right)$

Mass conservation: $\dfrac{d}{dx}\left(\rho V_x \right) = 0 \Rightarrow \dot{m}'' = \left(\rho V_x \right) = const$ (1)

Species conservation:

Fuel:

$$\rho V_x \frac{dY_F}{dx} = \rho D \frac{d^2 Y_F}{dx^2} + \dot{m}_F''' \tag{2}$$

Oxidizer:

$$\rho V_x \frac{dY_{ox}}{dx} = \rho D \frac{d^2 Y_{ox}}{dx^2} + \dot{m}_{ox}''' \tag{3}$$

Product:

$$\rho V_x \frac{dY_P}{dx} = \rho D \frac{d^2 Y_P}{dx^2} + \dot{m}_p''' \tag{4}$$

Energy equation:

$$\rho V_x C_p \frac{dT}{dx} = k_g \frac{d^2 T}{dx^2} - \sum \hat{h}_{f,i}^0 \dot{m}_i''' \tag{5}$$

Global reaction mechanism:

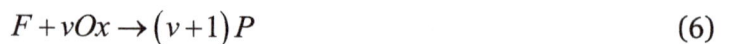

$$F + vOx \rightarrow \left(v + 1 \right) P \tag{6}$$

Heat release due to chemical reaction:

$$\sum \hat{h}_{f,i}^0 \dot{m}_i''' = \left[\hat{h}_{f,F}^0 \dot{m}_F''' + v\hat{h}_{f,Ox}^0 \dot{m}_F''' - \left(v + 1 \right) \hat{h}_{f,p}^0 \dot{m}_F''' \right] = \dot{m}_F''' \Delta \hat{H}_c \tag{7}$$

Now energy equation becomes

$$\dot{m}'' C_p \frac{dT}{dx} = k_g \frac{d^2T}{dx^2} - \dot{m}_F''' \Delta \hat{H}_c \tag{8}$$

Boundary conditions (For preheat zone)

$$x = -\infty; T = T_u; \frac{dT}{dx} = 0$$

$$x = +\infty; T = T_F; \frac{dT}{dx} = 0$$

Recasted energy equation for preheat zone,

$$\dot{m}'' C_p \frac{dT}{dx} = k_g \frac{d^2T}{dx^2} \tag{9}$$

Heat transfer due to conduction is balanced by convective heat transfer.

In the reaction zone,

$$x = -\infty; T = T_u; \frac{dT}{dx} = 0$$

$$x = -x_{ig}; T = T_{ig}$$

$$\left. \frac{dT}{dx} \right|_{ig} = \frac{\dot{m}'' C_p}{k_g} \left(T_{ig} - T_u \right) \tag{10}$$

$$k_g \frac{d^2T}{dx^2} = \dot{m}_F''' \Delta H_c \tag{11}$$

Rewriting, E.g. (11)

$$dT \frac{d}{dx} \left(\frac{dT}{dx} \right) = \frac{\Delta H_c}{k_g} \dot{m}_F''' dT \tag{12}$$

$$\left(\frac{d}{dx} \right)_{ig} = \left[\frac{2\hat{H}_c}{k_g} \int_{T_{ig}}^{T_F} \dot{m}_F''' dT \right]^{0.5} \tag{13}$$

Combining equations (10) and (13),

$$\frac{\dot{m}'' C_p}{k_g} \left(T_{ig} - T_u \right) = \left[\frac{2\Delta \hat{H}_c}{k_g} \int_{T_{ig}}^{T_F} \dot{m}_F''' dT \right]^{0.5} \tag{14}$$

$$\Rightarrow \dot{m}'' = \frac{k_g}{C_p} \frac{1}{(T_{ig} - T_u)} = \left[\frac{2\Delta\hat{H}_c}{k_g} \int_{T_{ig}}^{T_F} \dot{m}_F''' dT \right]^{0.5} \tag{15}$$

Also,

$$\Rightarrow \dot{m}'' = \rho_u S_L \dots \tag{16}$$

Combining equations (15) and (16),

$$S_L = \frac{k_g}{\rho_u C_p} \frac{4}{3(T_F - T_u)} = \left[\frac{2\Delta H_c}{k_g} \int_{T_{ig}}^{T_F} \dot{m}_F''' dT \right]^{0.5} \tag{17}$$

Mean fuel burning rate per unit volume,

$$\overline{\dot{m}}_F'' = \frac{1}{(T_F - T_u)} \left[\int_{T_u}^{T_F} \dot{m}_F''' dT \right] \tag{18}$$

Mean fuel burning rate can also be expressed as,

$$\overline{\dot{m}}_F'' = MW_F A_f C_F^{m_1} C_{ox}^{m_2} e^{-E/R_u T} \tag{19}$$

Expression for burning velocity becomes,

$$S_L = \left[\frac{32\alpha}{9\rho_u} (v+1) \overline{\dot{m}}_F'' \right]^{0.5}$$

Flame Thickness

Flame Thickness (d_L)	Ratio of maximum temperature difference to the temperature difference at the inflection point

$$\delta_L \equiv \frac{(T_F - T_u)}{(dT/dx)_{ig}}$$

Ignition temperature can be approximated as

$$T_{ig} = 0.75 T_F + 0.25 T_u$$

The temperature gradient at the flame surface is

$$\left. \frac{dT}{dx} \right|_{ig} = \frac{3}{4} \frac{\dot{m}'' C_p}{k_g} (T_F - T_u)$$

Flame thickness:

$$\delta_L \equiv \frac{4}{3}\frac{k_g}{C_p}\frac{1}{\dot{m}''} \quad \delta_L = \frac{4}{3}\frac{\alpha}{S_L} \text{ Where,} \quad \alpha = k_g / \rho C_p$$

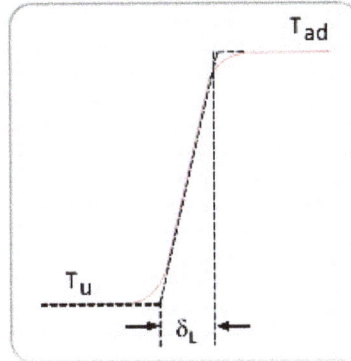

Burning Velocity Measurement Methods

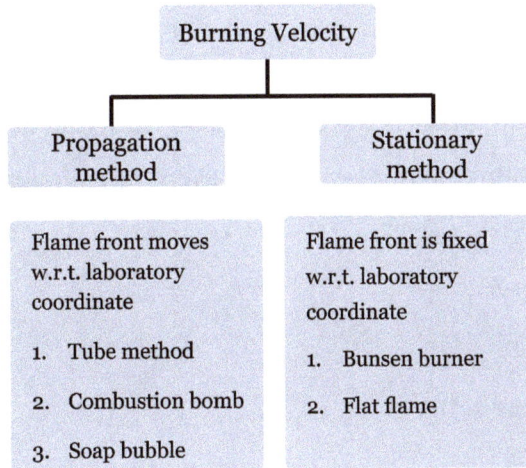

Flame front visualization

- Luminous photography

- Shadowgraph photography

- Schlieren Photography

Luminous	• Luminous part occurs at the burnt side • Flame speed w.r.t. unburnt gas is needed
Shadowgraph	• Corresponds to 2nd derivative of density • Closer to inflection point in temp. profile
Schlieren	• Captures maximum density gradient • Closer to unburnt mixture- preferred one

Tube Method

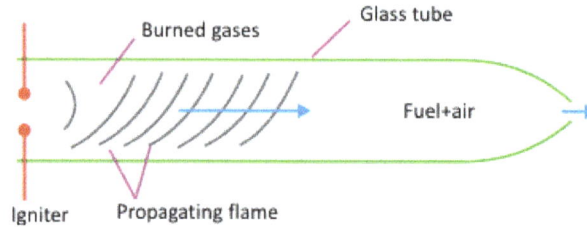

Procedure:

- Combustible mixture is filled in the tube

- On ignition at one end, the flame propagates through the tube

Features:

- Inner dia of tube should be greater than the quenching diameter

- The flame is planar in the beginning and curved towards downstream, due to buoyancy

- Natural convection distorts the planar flame front due to difference in densities

- Friction at the tube wall is also a reason for parabolic shape of the flame

The burning velocity is given by $S_L = \left(V_F - V_g \right) A_t / A_F$

V_F : Flame front velocity

V_g : Unburnt gas velocity

A_t : Cross-sectional area of tube

A_F : Flame surface area

Combustion Bomb Method

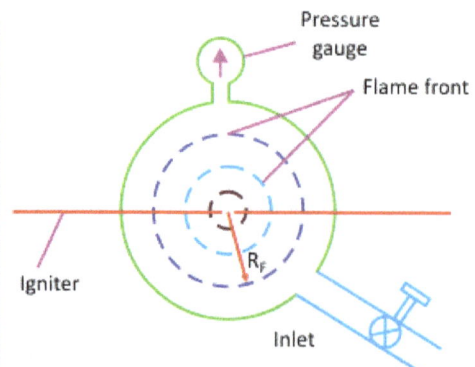

- Combustible mixture is ignited at the center of spherical vessel (constant volume)

- Flame propagates towards the wall

- Pressure and temperature increases due to adiabetic compression

- If the flame front radius is known, then

$$S_L = \frac{dR_f}{dt} \frac{\rho_b}{\rho_u} = V_F \frac{\rho_b}{\rho_u}$$

R_f : instantaneous radius of spherical flame

ρ_b : density of gas mixture at burnt state

ρ_u : density of unburnt mixture

Assumptions:

- Effect of flame front thickness and curvature are negligibly small
- Pressure at any instant is uniform throughout the vessel
- No heat loss including radiation
- Chemical equilibrium is achieved behind the flame

Soap Bubble Method

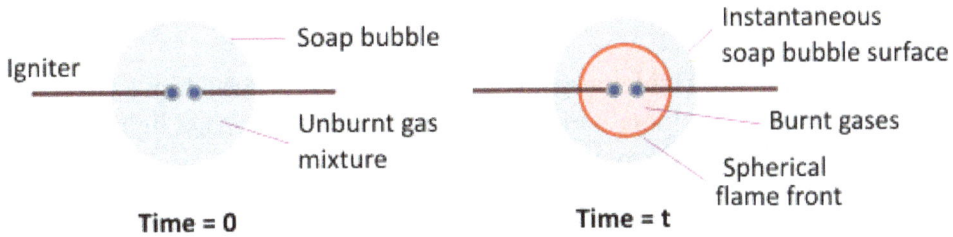

Igniter — Soap bubble

Unburnt gas mixture

Time = 0

Instantaneous soap bubble surface

Burnt gases

Spherical flame front

Time = t

- Homogeneous fuel-oxidizer mixture is confined in a soap bubble.
- On ignition at the center, spherical flame propagates.
- Pressure of burnt gas remains constant as the flame propagates.
- Assuming flame to be spherical and pressure remaining constant, a mass flux balance yields,

$$S_L A \rho_u = V_F A \rho_b \qquad S_L = V_F \left(\frac{\rho_b}{\rho_u} \right) = V_F \left(\frac{T_u}{T_b} \right)$$

- V_F : flame front velocity, ρ_b: density of gas mixture at burned state, ρ_u: unburnt gas density
- This method cannot be used for dry mixture.
- Flame front may not retain the spherical shape.
- Flame front would not be smooth for fast flames.
- Heat loss to electrode and ambient environment incurs error.

Stationary Flame Method (Bunsen Burner)

- The gas burns at the exit of the tube and a conical flame is established
- For flame to be stationary, the local burning velocity must be equal to the local flow velocity
- Flame shape will be influenced by the exit velocity profile and heat loss to the tube wall
- Lengthy tube ensures fully developed flow
- For a stationary flame,

    mass balance provides expression for S_L

$$S_L = V_t \left(\frac{A_t}{A_F} \right)$$

V_t : average flow velocity in tube, A_t : tube cross sectional area, A_F : conical surface area of flame
This method is known as area method

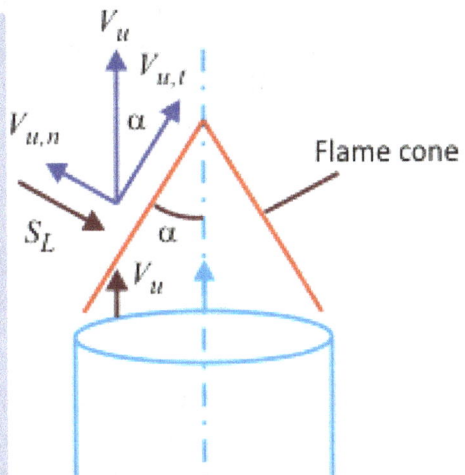

V_u

$V_{u,t}$

α

$V_{u,n}$

S_L

α

V_u

Flame cone

This kind of burner is suitable for mixture having low burning velocity (≤ 15 cm/s)

Bunsen Burner

A Bunsen burner, named after Robert Bunsen, is a common piece of laboratory equipment that produces a single open gas flame, which is used for heating, sterilization, and combustion.

The gas can be natural gas (which is mainly methane) or a liquefied petroleum gas, such as propane, butane, or a mixture of both.

History

In 1852 the University of Heidelberg hired Bunsen and promised him a new laboratory building. The city of Heidelberg had begun to install coal-gas street lighting, and so the university laid gas lines to the new laboratory.

The designers of the building intended to use the gas not just for illumination, but also in burners for laboratory operations. For any burner lamp, it was desirable to maximize the temperature and minimize luminosity. However, existing laboratory burner lamps left much to be desired not just in terms of the heat of the flame, but also regarding economy and simplicity.

While the building was still under construction in late 1854 Bunsen suggested certain design principles to the university's mechanic, Peter Desaga, and asked him to construct a prototype. (Similar principles had been used in an earlier burner design by Michael Faraday as well as in a device patented in 1856 by the gas engineer R. W. Elsner) The Bunsen/Desaga design succeeded in generating a hot, sootless, non-luminous flame by mixing the gas with air in a controlled fashion before combustion. Desaga created adjustable slits for air at the bottom of the cylindrical burner, with the flame igniting at the top. By the time the building opened early in 1855 Desaga had made fifty of the burners for Bunsen's students. Two years later Bunsen published a description, and many of his colleagues soon adopted the design. Bunsen burners are now used in laboratories all around the world.

Operation

The device in use today safely burns a continuous stream of a flammable gas such as natural gas (which is principally methane) or a liquefied petroleum gas such as propane, butane, or a mixture of both.

The hose barb is connected to a gas nozzle on the laboratory bench with rubber tubing. Most laboratory benches are equipped with multiple gas nozzles connected to a central gas source, as well as vacuum, nitrogen, and steam nozzles. The gas then flows up through the base through a small hole at the bottom of the barrel and is directed upward. There are open slots in the side of the tube bottom to admit air into the stream via the venturi effect, and the gas burns at the top of the tube once ignited by a flame or spark. The most common methods of lighting the burner are using a match or a spark lighter.

The amount of air mixed with the gas stream affects the completeness of the combustion reaction. Less air yields an incomplete and thus cooler reaction, while a gas stream well mixed with air provides oxygen in an equimolar amount and thus a complete and hotter reaction. The air flow can be controlled by opening or closing the slot openings at the base of the barrel, similar in function to the choke in a carburettor.

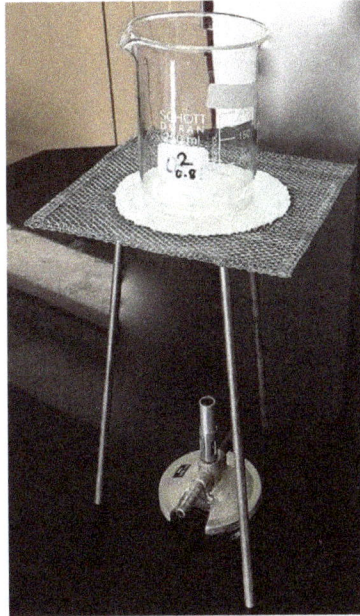

A Bunsen burner situated below a tripod

If the collar at the bottom of the tube is adjusted so more air can mix with the gas before combustion, the flame will burn hotter, appearing blue as a result. If the holes are closed, the gas will only mix with ambient air at the point of combustion, that is, only after it has exited the tube at the top. This reduced mixing produces an incomplete reaction, producing a cooler but brighter yellow which is often called the "safety flame" or "luminous flame". The yellow flame is luminous due to small soot particles in the flame which are heated to incandescence. The yellow flame is considered "dirty" because it leaves a layer of carbon on whatever it is heating. When the burner is regulated to produce a hot, blue flame it can be nearly invisible against some backgrounds. The hottest part of the flame is the tip of the inner flame, while the coolest is the whole inner flame. Increasing the amount of fuel gas flow through the tube by opening the needle valve will increase the size of the flame. However, unless the airflow is adjusted as well, the flame temperature will decrease because an increased amount of gas is now mixed with the same amount of air, starving the flame of oxygen.

Generally, the burner is placed underneath a laboratory tripod, which supports a beaker or other container. The burner will often be placed on a suitable heatproof mat to protect the laboratory bench surface.

Variants

Other burners based on the same principle exist. The most important alternatives to the Bunsen burner are:

- Teclu burner – The lower part of its tube is conical, with a round screw nut below its base. The gap, set by the distance between the nut and the end of the tube, regulates the influx of the air in a way similar to the open slots of the Bunsen burner. The Teclu burner provides better mixing of air and fuel and can achieve higher flame temperatures than the Bunsen burner.

- Meker burner – The lower part of its tube has more openings with larger total cross-section, admitting more air and facilitating better mixing of air and gas. The tube is wider and its top is covered with a wire grid. The grid separates the flame into an array of smaller flames with a common external envelope, and also prevents flashback to the bottom of the tube, which is a risk at high air-to-fuel ratios and limits the maximum rate of air intake in a conventional Bunsen burner. Flame temperatures of up to 1,100–1,200 °C (2,000–2,200 °F) are achievable if properly used. The flame also burns without noise, unlike the Bunsen or Teclu burners.

- Tirrill burner – The base of the burner has a needle valve which allows the regulation of gas intake directly from the Burner, rather than from the gas source. Maximum temperature of flame can reach 1560 °C.

Flat Flame Burner

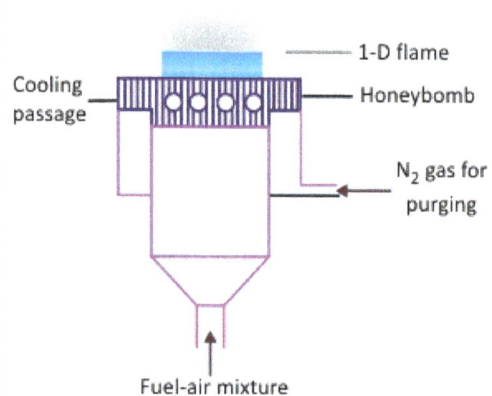

- 1D steady flame can be easily established
- 1D velocity profile is obtained using honey comb or a metal porous plug
- Invert gas curtain can be used to prevent diffusion of atmospheric air
- Grid can be used to stabilize the flame
- Area of flat flame- measured by photo
- Flame area is same for visible, shadow schlieren photographs
- Simplest and most accurate of all the methods discussed so far

1-D flame
Honeybomb
Cooling passage
N_2 gas for purging
Fuel-air mixture

This kind of burner is suitable for mixture having low burning velocity ($\leq$15 cm/s)

Effect of Equivalence Ratio on S_L

- For hydrocarbon fuels, $S_L \approx 25 - 50$ cm/s
- Peak burning velocity occurs at $\phi = 1.05$
- For hydrogen-air mixture,
  - Peak burning velocity occurs at $\phi = 1.8$
- Since S_L MW^{-1}; Maximum S_L occurs under fuel-rich conditions
- For CO-air mixture also maximum S_L occurs under fuel-rich conditions
- CO-air burning velocity data is not understood due to complex chemical kinetics
- The peak burning velocity for H_2, C_2H_2 and CO are quite different and cannot be explained in terms of molecular weight

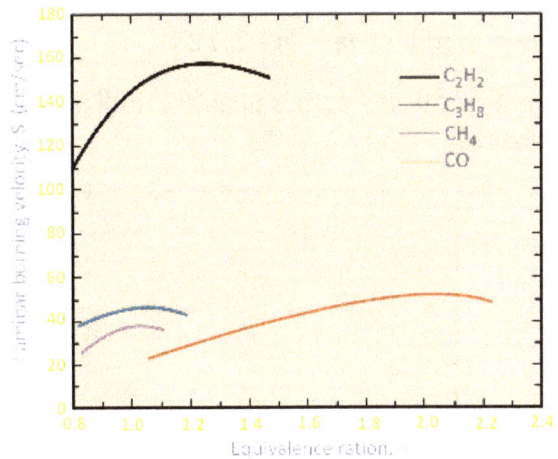

Effect of Oxygen Concentration on S_L

For various combinations of O_2 and N_2, there is a drastic increase in S_L.

For methane, S_L is increased by 10 times

For propane, S_L is increased by 7.5 times

For CO, S_L is increased by 2.4 times

What is the reason for this change?

Higher the oxygen level, higher will be the adiabatic flame temperature

Thus higher burning velocity

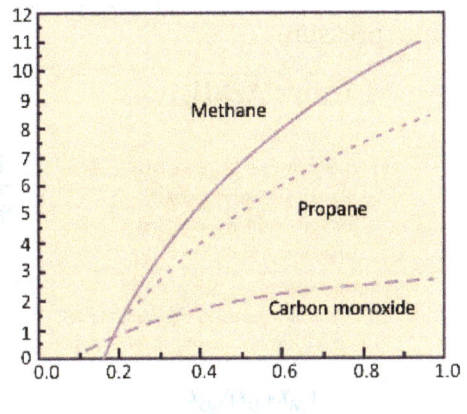

Effect of Initial Pressure and Temperature on S_L

$$S_L \propto P^{(n-2)/2}$$

This relation is supported by experimental data

n: Overall order of global chemical reaction

$n < 2$ for HC flames with $S_L < 50 \, cm/s$

$n = 2$ for HC flames with $S_L \approx 50 - 100 \, cm/s$

$n > 2$ for HC flames with $S_L > 100\ cm/s$

Pressure index, $m = (n-2)/2$

For $S_L < 50\ cm/s$, m is negative, indicating burning velocity increases with decreasing pressure

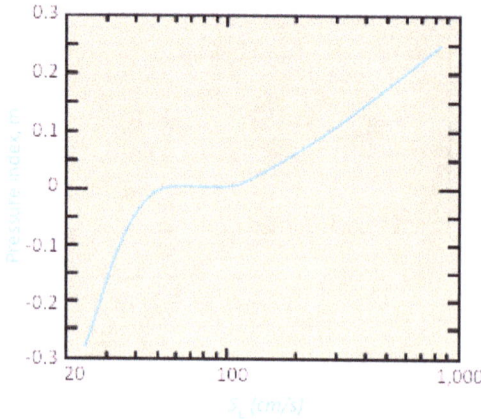

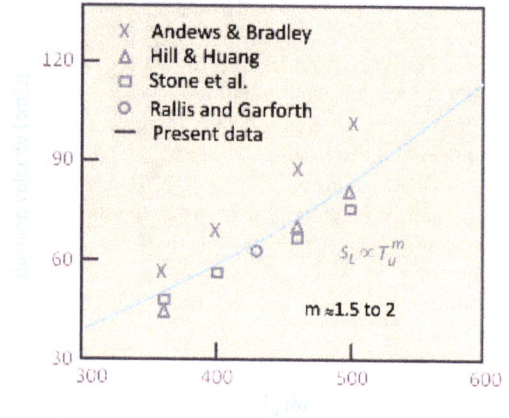

For $S_L \approx 50-100 cm/s$, m is constant, indicating burning velocity is constant

For $S_L > 100 cm/s$, m is positive, indicating burning velocity decreases with decrease in initial pressure

Effect of Inert Additives

Inert gas additives produce the following effects
1. Reduction of burning velocity
2. Narrowing of flammability limits
3. Shifting of S_L peak

Why is there shift in S_L peak?

Addition of inert gas changes the ratio of thermal conductivity and sp. heat of the mixture, resulting in change in S_L.

Burning velocity with He addition is higher than Ar and N2. Why?

Lower MW will lead to higher burning velocity MW of He is much lower than Ar and hence higher S_L

Ar and N2 have almost same thermal diffusivity Hence for same heat release, Ar attains higher flame temperature ➡ higher S_L

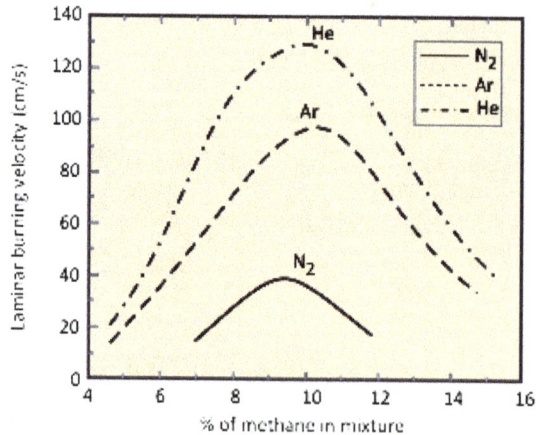

Clingman et al 1995

Flammability Limit

Mixtures of dispersed combustible materials (such as gaseous or vaporised fuels, and some dusts) and air will burn only if the fuel concentration lies within well-defined lower and upper bounds determined experimentally, referred to as flammability limits or explosive limits. Combustion can range in violence from deflagration, through detonation, to explosion.

Limits vary with temperature and pressure, but are normally expressed in terms of volume percentage at 25°C and atmospheric pressure. These limits are relevant both to producing and optimising explosion or combustion, as in an engine, or to preventing it, as in uncontrolled explosions of build-ups of combustible gas or dust. Attaining the best combustible or explosive mixture of a fuel and air (the stoichiometric proportion) is important in internal combustion engines such as gasoline or diesel engines.

The standard reference work is that by Zabetakis using an apparatus developed by the United States Bureau of Mines.

Violence of Combustion

Combustion can vary in degree of violence. A deflagration is a propagation of a combustion zone at a velocity less than the speed of sound in the unreacted medium. A detonation is a propagation of a combustion zone at a velocity greater than the speed of sound in the unreacted medium. An explosion is the bursting or rupture of an enclosure or container due to the development of internal pressure from a deflagration or detonation as defined in NFPA 69.

Limits

Lower Explosive Limit

Lower explosive limit (LEL): The lowest concentration (percentage) of a gas or a vapor in air capable of producing a flash of fire in presence of an ignition source (arc, flame, heat). The term is considered by many safety professionals to be the same as the lower flammable limit (LFL). At a concentration in air lower than the LEL, gas mixtures are "too lean" to burn. Methane gas has an LEL of 5.0%. If the atmosphere has less than 5.0% methane, an explosion cannot occur even if a source of ignition is present.

Percentage reading on combustible air monitors should not be confused with the LEL concentrations. Explosimeters designed and calibrated to a specific gas may show the relative concentration of the atmosphere to the LEL—the LEL being 100%. A 5% displayed LEL reading for methane, for example, would be equivalent to 5% multiplied by 5.0%, or approximately 0.25% methane by volume at 20 degrees C. Control of the explosion hazard is usually achieved by sufficient natural or mechanical ventilation, to limit the concentration of flammable gases or vapors to a maximum level of 25% of their lower explosive or flammable limit.

Upper Explosive Limit

Upper explosive limit (UEL): Highest concentration (percentage) of a gas or a vapor in air capable of producing a flash of fire in presence of an ignition source (arc, flame, heat). Concentrations higher than UFL or UEL are "too rich" to burn.

Influence of Temperature, Pressure and Composition

Flammability limits of mixtures of several combustible gases can be calculated using Le Chatelier's mixing rule for combustible volume fractions x_i:

$$LEL_{mix} = \frac{1}{\sum \dfrac{x_i}{LEL_i}}$$

and similar for UEL.

Temperature, pressure, and the concentration of the oxidizer also influences flammability limits. Higher temperature or pressure, as well as higher concentration of the oxidizer (primarily oxygen in air), results in lower LFL and higher UFL, hence the gas mixture will be easier to explode. The effect of pressure is very small at pressures below 10 millibar and difficult to predict, since it has only been studied in internal combustion engines with a turbocharger.

Usually atmospheric air supplies the oxygen for combustion, and limits assume the normal concentration of oxygen in air. Oxygen-enriched atmospheres enhance combustion, lowering the LFL and increasing the UFL, and vice versa; an atmosphere devoid of an oxidizer is neither flammable nor explosive for any fuel concentration. Significantly increasing the fraction of inert gases in an air mixture, at the expense of oxygen, increases the LFL and decreases the UFL.

Controlling Explosive Atmospheres

Gas and Vapor

Controlling gas and vapor concentrations outside the explosive limits is a major consideration in occupational safety and health. Methods used to control the concentration of a potentially explosive gas or vapor include use of sweep gas, an unreactive gas such as nitrogen or argon to dilute the explosive gas before coming in contact with air. Use of scrubbers or adsorption resins to remove explosive gases before release are also common. Gases can also be maintained safely at concentrations above the UEL, although a breach in the storage container can lead to explosive conditions or intense fires.

Dusts

Dusts also have upper and lower explosion limits, though the upper limits are hard to measure and of little practical importance. Lower explosive limits for many organic materials are in the range of 10–50 g/m³, which is much higher than the limits set for health reasons, as is the case for the LEL of many gases and vapours. Dust clouds of this concentration are hard to see through for more than a short distance, and normally only exist inside process equipment.

Explosion limits also depend on the particle size of the dust involved, and are not intrinsic properties of the material. In addition, a concentration above the LEL can be created suddenly from settled dust accumulations, so management by routine monitoring, as is done with gases and vapours, is of no value. The preferred method of managing combustible dust is by preventing accumulations of settled dust through process enclosure, ventilation, and surface cleaning. However, lower explosion limits may be relevant to plant design.

Volatile Liquids

Situations caused by evaporation of flammable liquids into the air-filled void volume of a container may be limited by flexible container volume or by using an immicsible fluid to fill the void volume. Hydraulic tankers use displacement of water when filling a tank with petroleum.

Ignition

Ignition

Rate of heat liberation near the ignition zone > rate of heat loss by conduction

Energy generated in flame = Ensible enthalpy × Mass

$$MIE = C_p \left(T_F - T_u \right) \left[\left(\frac{\pi}{4} d_q^2 \right) S_L \rho u \right]$$

Substituting quenching diameter

$$MIE = 2\pi C \rho_u C_p \left(T_F - T_u \right) S_L^3$$

Substituting flame thickness

$$MIE = \frac{128\pi}{27} \frac{C \left(T_F - T_u \right) \alpha^2 k_g}{S_L^3}$$

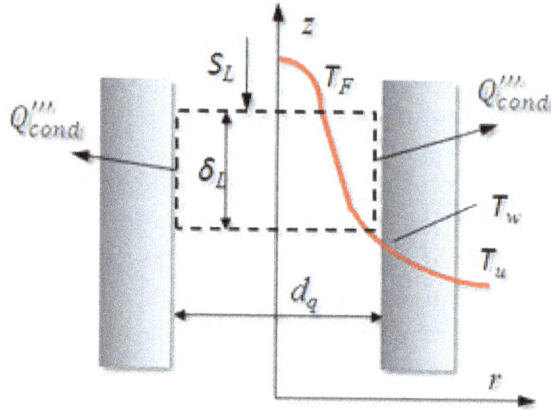

Dependence on pressure

$$MIE \approx \frac{1}{\rho_u^2 S_L^3} \approx p^{-(3n/2-1)}$$

Fuel-Air	Fuel-Air
Methane-air	Methane-air
Ethane-air	Ethane-air
Butane-air	Butane-air
Acetylene-air	Acetylene-air
CO-air	CO-air
Hydrogen-air	Hydrogen-air

Flame Stabilization

Local gas flow velocity = Local burning velocity

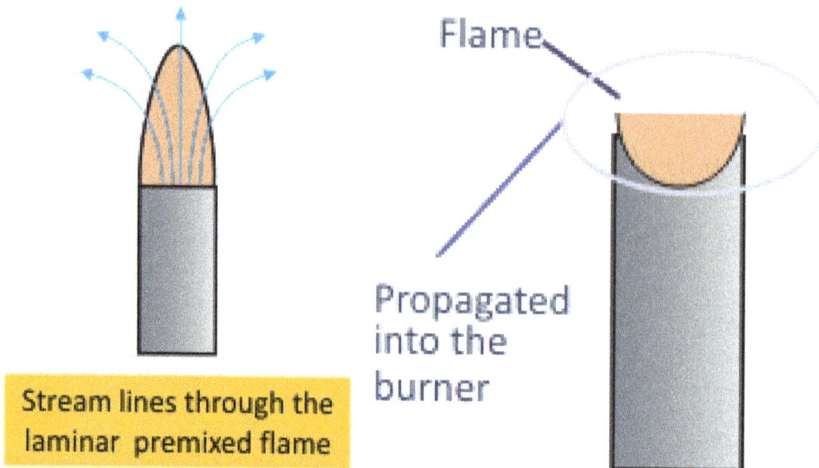

Stream lines through the laminar premixed flame

Flame

Propagated into the burner

Flame Stabilization by Burner Rim

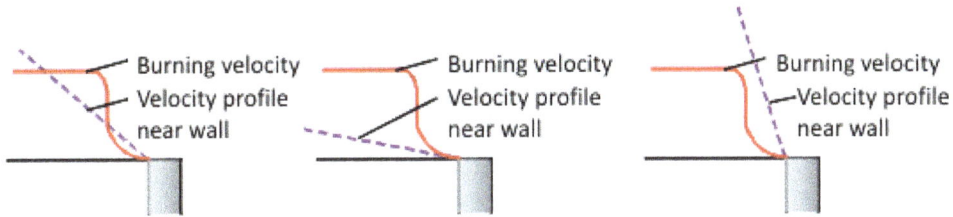

Stability of flame front near the rim of Bunsen burner

- At burner rim, flow velocity ≈ burning velocity: flame likely to get stabilized here
- Heat loss and radical loss at the burner rim is the cause for flame stabilization
- Laminar velocity profile can be obtained at low Reynolds number
- When flow velocity (V_F) < burning velocity (S_L),
  - Flame enters burner, leading to flash back
  - At the critical condition, the velocity gradient (g_F) is $g_F = \lim_{r \to R} \left(-\dfrac{dV}{dr} \right)$

The parabolic velocity profile is $V = n(R^2 - r^2)$; R- tube radius;

$$n = -\frac{1}{4\mu}(\Delta P / L)$$

ΔP is pressure difference across tube length, μ is the fluid viscosity

At flash back, $g_b = 8 \dfrac{V_{av}}{d}$

Turbulent Premixed Flame

Turbulence in flame	• Affects flame propagation rate • Does not alter chemistry • Mixing occurs due to random motion of eddies

- Turbulent flames are chaotic in nature
- Instantaneous flame front is highly convoluted
- Actual position of reaction zone moves rapidly in space w.r.t. time
- This makes flame to appear thick
- Turbulent flame brush: Virtual turbulent flame thickness
- Laminar flamelets: Instantaneous reaction zone

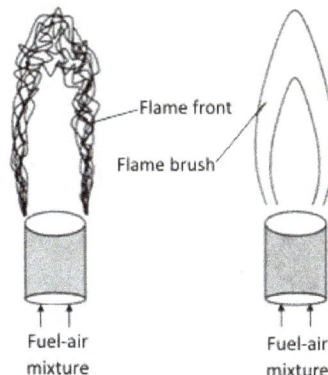

Turbulent Flame Regimes

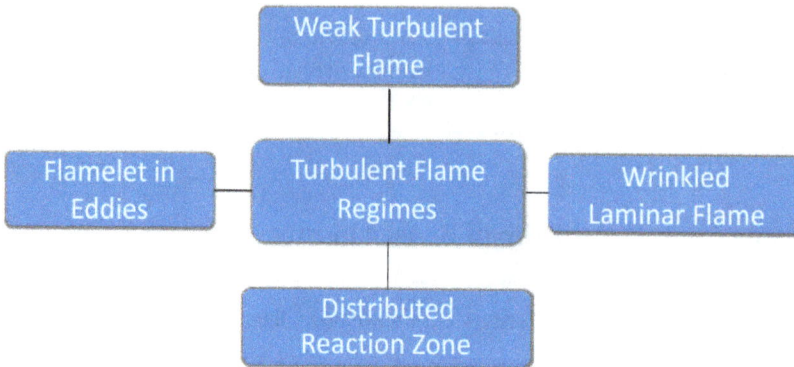

```
              ┌─────────────────┐
              │ Weak Turbulent  │
              │     Flame       │
              └─────────────────┘
                       │
┌──────────────┐  ┌──────────────────┐  ┌──────────────┐
│ Flamelet in  │──│ Turbulent Flame  │──│   Wrinkled   │
│    Eddies    │  │     Regimes      │  │ Laminar Flame│
└──────────────┘  └──────────────────┘  └──────────────┘
                       │
              ┌─────────────────┐
              │   Distributed   │
              │ Reaction Zone   │
              └─────────────────┘
```

Reynolds number for turbulent flame:

$$\mathrm{Re}_l = \frac{V'_{rms} l_0}{v}$$

Chemical reaction time:

$$\tau_{ch} = \frac{\delta_L}{S_L}$$

Chemical reaction time:

$$\tau_t = \frac{l_0}{V'_{rms}}$$

Damkohler number

$$Da = \frac{t_m}{t_{ch}} = \frac{l_0 / V'_{rms}}{\delta_L / S_L}$$

If Da >> 1, fast chemistry regime

If Da <<1, fast mixing regime

The Borghi Diagram

Borghi Diagram

- The plot of Da against Re_1 on a log-log scale
- Depicts various regimes of turbulent flames

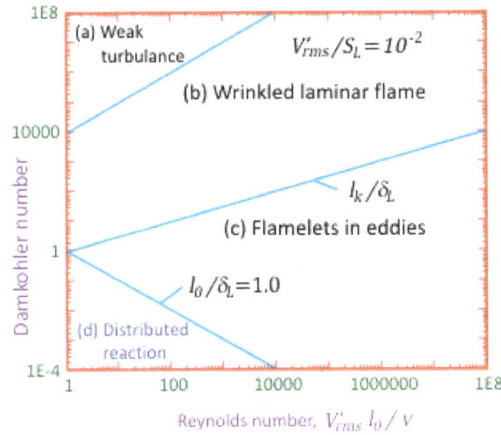

Weak turbulent flame

- Upper region of the Borghi diagram

Wrinkled laminar flame

- Region between $V'_{rms} / S_L = 10^{-2}$ and l_k / δ_L

- Chemical reaction takes place in a thin zone

Flamelets in eddies

- Region between upper bold line $l_k / \delta_L = 1$ and $l_0 / \delta_L = 1$

Distributed reaction regime

- Region below $l_0 / \delta_L = 1$

- Reaction sheets are distributed in the turbulent flame surface $\left(Da < 1 \right)$

- This type of combustion can be established in a well stirred reactor.

Turbulent Burning Velocity

Turbulent burning Velocity (S_T)	<ul><li>Depends on characteristics of fluid flow</li><li>Velocity at which unburnt mixture enters the flame zone</li><li>Difficult to measure velocity of unburnt gas near the turbulent flame</li></ul>

How to measure (ST) ?

From the reactant flow rate

$$S_T = \frac{\dot{m}}{\overline{A} p_u}$$

Turbulent Burning Velocity

$$\frac{S_T}{S_L} = \left(\frac{\alpha_T}{\alpha_L}\right)^{0.5}$$

$\dot{m}$ is the reactant flow rate

$\overline{A}$ is the time average flame surface area

is the density of unburnt gas

Weak Turbulent
Flame

- Extension of laminar flame (low turbulence level)
- Smooth flame
- Turbulence scale $\approx$ Laminar flame thickness
- $S_T > S_L$.................Why ? increase in thermal diffusion

Wrinkled Laminar Flame

- Flamelets in flame surface propagate at laminar burning velocity
- Turbulence only causes wrinkling of flame

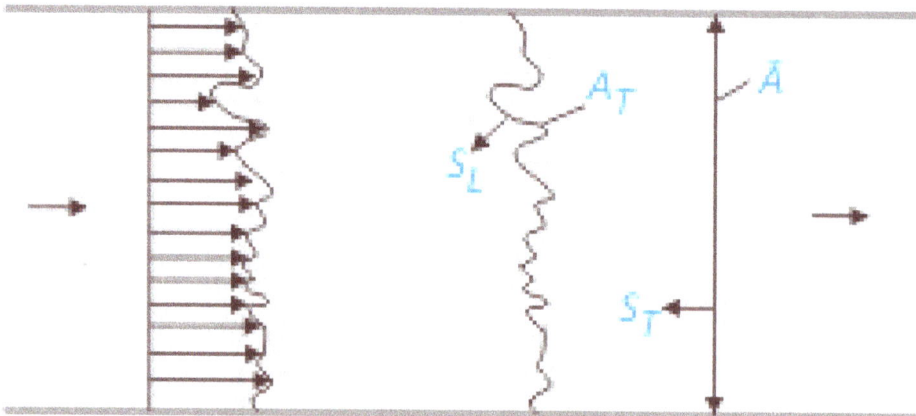

Turbulent burning velocity is given by,

$$m = \rho_u \overline{A} S_T = \rho_u A_{F1} S_L \Rightarrow \frac{S_T}{S_L} = \frac{A_{F1}}{\overline{A}}$$

According to Damkohler, for constant laminar burning velocity

$$\frac{A_{F1}}{\overline{A}} = \frac{V_u}{S_L}$$

According to Klimov,

$$\frac{S_T}{S_L} = 3.5 \left(\frac{V'_{rms}}{S_L} \right)^{0.7}$$

Similarly for turbulent flame,

$$\frac{A_w}{A} = \frac{A'_{rms}}{S_L}$$

$$\frac{S_T}{S_L} = \frac{\overline{A} + A_w}{\overline{A}} = \left(1 + \frac{V'_{rms}}{S_L} \right)$$

According to Calvin and William,

$$\frac{S_T}{S_L} = \left[0.5 \left\{ 1 + \left(1 + \frac{8CV'^2_{rms}}{S_L^2} \right)^{0.5} \right\} \right]^{1/2}$$

Distributed Reaction

- For high intensities of wrinkling, distinct regime with small pockets of reactants are formed.
- In Borghi's diagram, this lies in the regime where integral length scale and Damkohler number are less than unity.
- Quite difficult to occur in practical devices.
- In laboratory, such situation can be created by using stirred reactor.
- Chemical reactions are not completed in reaction zone; rather occur in post-flame region.
- This regime needs more understanding.

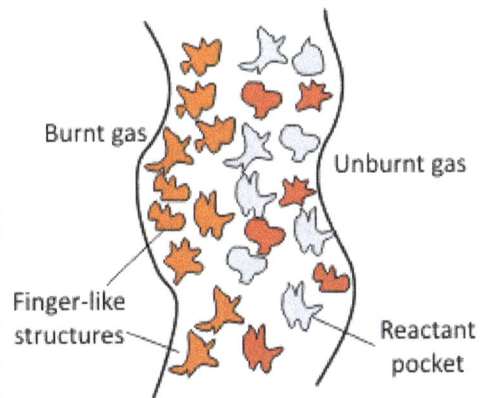

Turbulent burning velocity is given by,

$$S_T = 6.4 V'_{rms} \left(\frac{\overline{V}}{V'_{rms}} \right)^{3/4}$$

Flamelet in Eddies

Regime in Borghi's diagram	Region between upper bold line $(I_k/\delta_L=1)$ and lower bold line $(I_o/\delta_L=1)$
Reaction zone	parcels of unburnt gas and fully burnt gases

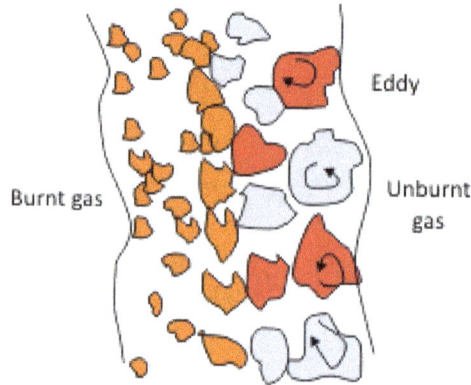

Fuel mass burning rate,

$$\dot{m}_F''' = -\rho C_F Y'_{F,rms} \varepsilon_0 / ke_t$$

Typically, $C_F = 1$; $Y'_{F,rms}$ is root mean square of fluctuating fuel mass fraction, ke_t is the turbulent kinetic energy per unit.

$$\dot{m}_F''' = -\rho C_F Y'_{F,rms} V_{rms} / l_o$$

References

- Kerstein, Alan R. (1988-01-01). "Field equation for interface propagation in an unsteady homogeneous flow field". Physical Review A. 37 (7): 2728–2731. doi:10.1103/PhysRevA.37.2728

- Lewis, Bernard; Elbe, Guenther von (2012). Combustion, Flames and Explosions of Gases. Elsevier. ISBN 9780323138024

- Lockemann, G. (1956). "The Centenary of the Bunsen Burner". J. Chem. Ed. 33: 20–21. Bibcode:1956JChEd..33...20L. doi:10.1021/ed033p20

- Ihde, Aaron John (1984). The development of modern chemistry. Courier Dover Publications. pp. 233–236. ISBN 978-0-486-64235-2

- Bradley, D (2009-06-25). "Combustion and the design of future engine fuels". Proceedings of the Institution of Mechanical Engineers, Part C: Journal of Mechanical Engineering Science. 223 (12): 2751–2765. doi:10.1243/09544062jmes1519

- Kohn, Moritz (1950). "Remarks on the history of laboratory burners". J. Chem. Educ. 27 (9): 514. Bibcode:1950JChEd..27..514K. doi:10.1021/ed027p514

- Partha, Mandal Pratim & Mandal, B. (2002-01-01). A Text Book of Homoeopathic Pharmacy. Kolkata, India: New Central Book Agency. p. 46. ISBN 978-81-7381-009-1

- Teclu, Nicolae (1892). "Ein neuer Laboratoriums-Brenner". J. Prakt. Chem. 45 (1): 281–286. doi:10.1002/prac.18920450127

6

Principles of Diffusion in Combustion

A diffusion flame can be formed when the oxidizer fuses with the fuel through the method of diffusion. Diffusion flame produces more soot and burns slower in comparison to pre-mixed flame. A few instances of diffusion flames are forest fire, solid fuel combustion, liquid fuel combustion, and candle flame. This chapter elucidates the principles of diffusion in combustion.

Diffusion Flame

In combustion, a diffusion flame is a flame in which the oxidizer combines with the fuel by diffusion. As a result, the flame speed is limited by the rate of diffusion. Diffusion flames tend to burn slower and to produce more soot than premixed flames because there may not be sufficient oxidizer for the reaction to go to completion, although there are some exceptions to the rule. The soot typically produced in a diffusion flame becomes incandescent from the heat of the flame and lends the flame its readily identifiable orange-yellow color. Diffusion flames tend to have a less-localized flame front than premixed flames.

In a diffusion flame, combustion takes place at the flame surface only, where the fuel meets oxygen in the right concentration - the interior of the flame contains unburnt fuel. This is opposite to combustion in a premixed flame. The fire breather's spurting of fuel (likely kerosene), combined with strong convection flows due to intense heat gives a turbulent diffusion flame.

The contexts for diffusion may vary somewhat. For instance, a candle uses the heat of the flame itself to vaporize its wax fuel and the oxidizer (oxygen) diffuses into the flame from the surrounding air, while a gaslight flame (or the safety flame of a bunsen burner) uses fuel already in the form of a vapor.

The common flame of a candle is a classic example of a diffusion flame. Its yellow color owing to the large amount of incandescent soot particles in the incomplete combustion reaction of the flame.

Diffusion flames are often studied in counter flow (also called opposed jet) burners. Their interest is due to possible application in the flamelet model for turbulent combustion. Furthermore they provide a convenient way to examine strained flames and flames with holes. These are also known under the name of "edge flames", characterized by a local extinction on their axis because of the high strain rates in the vicinity of the stagnation point.

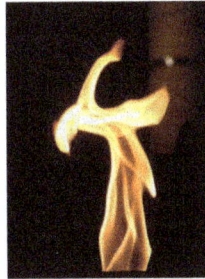

A nearly-turbulent diffusion flame

Diffusion flames have an entirely different appearance in a microgravity environment. There is no convection to carry the hot combustion products away from the fuel source, which results in a spherical flame front, such as in the candle seen here. This is a rare example of a diffusion flame which does not produce much soot and does not therefore have a typical yellow flame.

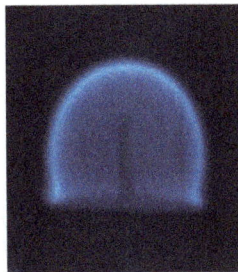

A candle in a microgravity environment. This is a rare example of a diffusion flame which does not produce much soot and does not therefore have a typical yellow flame.

Theoretical Analysis

Consider a 2D diffusion flame,

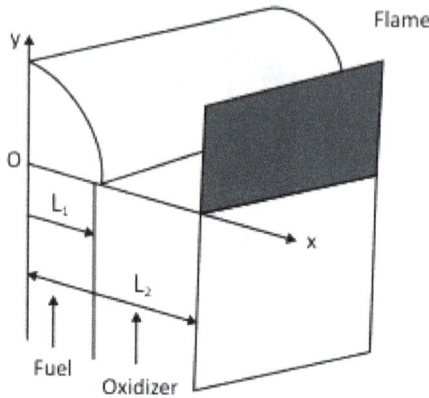

Assumptions:

i. 2D steady laminar inviscid flow.

ii. Velocity above the channel is constant everywhere, $V_x = 0$

iii. Fuel and oxidizer react in stoichiometric proportion at the flame surface with infinite reaction rate (Thin flame approximation).

iv. Binary diffusion between participating species.

v. Mass diffusion is along x-direction only.

vi. Unity Lewis number.

vii. Single step irreversible reaction.

viii. Radiation heat transfer is negligibly small.

ix. Constant thermophysical properties.

x. Mass diffusivity of both fuel and oxidizer are the same.

xi. Buoyancy force is neglected.

Conservation equations:

Mass conservation:

$$\frac{\partial(\rho V_x)}{\partial x} + \frac{\partial(\rho V_y)}{\partial y} = 0$$

Using assumption (ii), we can have,

$$\frac{\partial(\rho V_y)}{\partial y} = 0 \Rightarrow \rho V_y = const.$$

Axial momentum conservation:

$$\frac{\partial(\rho V_x V_y)}{\partial x} + \frac{\partial(\rho V_y V_y)}{\partial y} = \frac{\partial}{\partial x}\left(\mu \frac{dV_y}{dx}\right) + (\rho_\infty - \rho)g$$

The pressure gradient in the y direction is approximated as $\rho_\infty g$ - , which is known as Boussinesq approximation.

Species conservation equation:

$$\frac{\partial(\rho V_x Y_F)}{\partial x} + \frac{\partial(\rho V_y V_F)}{\partial y} = \frac{\partial}{\partial x}\left(\rho D_{12} \frac{\partial Y_F}{\partial x}\right) + \dot{m}_F'''$$

$$\frac{\partial(\rho V_x Y_{0x})}{\partial x} + \frac{\partial(\rho V_y V_{0x})}{\partial y} = \frac{\partial}{\partial x}\left(\rho D_{12} \frac{\partial Y_{0x}}{\partial x}\right) + \dot{m}_{0x}'''$$

Mass fraction of the product can be found from

$$Y_\rho = 1 - Y_F - Y_{0x}$$

By thin flame approximation,

$$V_y \frac{\partial Y_F}{\partial y} = D_{12} \frac{\partial^2 Y_F}{\partial x^2}; V_y \frac{\partial Y_{0x}}{\partial y} = D_{12} \frac{\partial^2 Y_{0x}}{\partial x^2}$$

Single step irreversible reaction,

$$F + vOx \rightarrow (v+1)P$$

Universal concentration variables,

$$\left.\frac{dY_F}{dx}\right|_{F^-} = -\frac{1}{v}\left.\frac{dY_{ox}}{dx}\right|_{F^+} \Rightarrow Y_R = Y_F = -\frac{Y_{ox}}{v}$$

Rate of fuel transport from the centre to the flame surface is equal to stoichiometric rate of oxidizer transport.

Let Y_R be the mass fraction of the reactant,

Instead of solving two equations (For fuel and oxidizer), we can solve a single equation as given below,

$$V_y \frac{\partial Y_R}{\partial y} = D_{12} \frac{\partial^2 Y_R}{\partial x^2}$$

This analysis is known as the Burke-Schumann's analysis

Above equation can be converted into a diffusion equation by substituting

$$y = V_y t$$

$$\frac{dY_R}{dt} = D_{12} \frac{d^2 Y_R}{dx^2}$$

Inner wall exists at $x = 0$ and outer wall at $x = L_2$

The initial and boundary conditions are as follows.

At $t = 0, Y_R = (Y_R)_0$ At $x = 0, \dfrac{dY_R}{dx} = 0; x = L_2, \dfrac{dY_R}{dx} = 0;$

Applying boundary conditions, we obtain a closed form series solution

$$\frac{Y_R}{(Y_R)_0} = \frac{(Y_F)_0}{(Y_R)_0} \frac{L_1}{L_2} - \frac{(Y_{0x})_0}{(Y_R)_0 v} \left(\frac{L_2 - L_1}{L_2} \right) + \frac{2}{\pi} \sum_{n=1}^{\infty} \frac{1}{n} \sin\left(\frac{n\pi L_1}{L_2} \right) \cos\left(\frac{n\pi x}{L_2} \right) \exp\left(\frac{-yn^2\pi^2 D_{12}}{vL_2^2} \right)$$

where, $Y_R / (Y_R)_0$ is the non-dimensional mass fraction of the reactant.

$$\Rightarrow (Y_R)_0 = (Y_F)_0 + (Y_{ox})_0 / v$$

The infinite series must have a constant value at the flame surface as given below

$$E = \frac{(Y_{ox})_0}{v(Y_R)_0} \left(\frac{L_2 - L_1}{L_2} \right) - \frac{(Y_F)_0}{(Y_R)_0} \frac{L_1}{L_2}$$

The series solution depends on L_1 / L_2, x / L_2 and ξ

At the burner rim, $\xi = 0$, the series constant (E)

becomes a square wave

$$E = \left(\frac{L_1}{L_2}\right) \text{ and } Y_R = \left(Y_F\right)_0 \text{ for } 0 < x \le L_1, \xi = 0$$

$$E = \left(\frac{L_1 - L_2}{L_2}\right) \text{ and } Y_R = -\frac{-\left(Y_{0x}\right)_0}{v} \text{ for } L_1 < x \le L_2, \xi = 0$$

When F/A ratio is stoichiometric E becomes zero.

$$\left(Y_F\right)_0 = \frac{\left(Y_{ox}\right)_0}{v}\left(\frac{L_2 - L_1}{L_2}\right)$$

Roper extended the Burke-Schumann model by varying the velocity to vary along the axial direction.

The flame height is given by,

$$h_{F,Roper} = \frac{\dot{V}_F\left(T_\infty / T_F\right)}{4\pi D_\infty \ln\left(1 + \frac{1}{v}\right)}\left(\frac{T_\infty}{T_{ad}}\right)^{0.67}$$

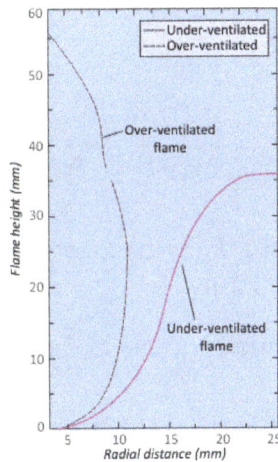

Phenomenological Analysis

Assumption:

The burning process is not being affected by the mixing rate between the fuel and the oxidizer.

The flame height is defined as the point along the z axis on which inter diffusion of the fuel and oxidizer is reached for the first time.

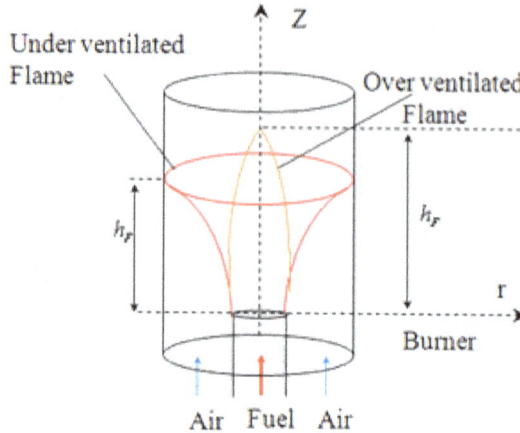

The time, t required for oxidizer element to reach axis of the jet is given by

$$t = h_F / V_Z$$

By the Einstein diffusion equation, we can have

$$\overline{r}^2 = 2D_{12}t$$

Where D12 is the fuel air diffusion coefficient, $\overline{r}^2$ is the average square displacement.

The expression for flame height is given by

$$h_F = \frac{V_z R^2}{2D_{12}}; \Rightarrow h_F = \frac{\rho V_z \pi R^2}{2\pi\rho D_{12}} = \frac{\rho \dot{V}}{2\pi k_g / C_p} = \frac{\dot{m}}{2\pi k_g / C_p}$$

where $\dot{V}$ is the volumetric flow rate, k_g is average thermal conductivity of mixture, C_P is the average of specific heat of mixture.

Flame height is independent of burner diameter for a particular volume flow rate.

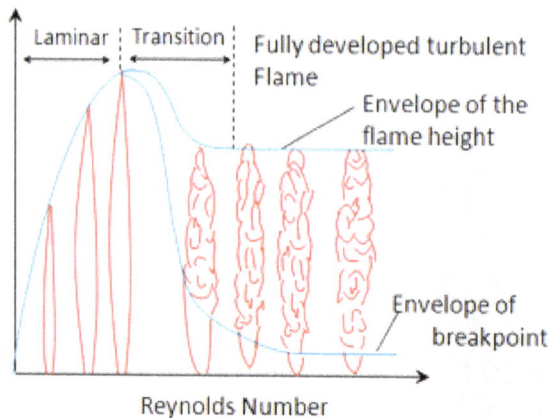

Transition Reynolds number	
Fuel-air	Retransition
Acetylene, C_2H_2	9500
Propane, C_3H_8	9500
City gas	3500
Carbon monoxide, CO	4800
Hydrogen, H_2	2000

- Laminar flame height increases linearly with nozzle velocity.

- The increase is observed until turbulent mixing occurs.

- This situation occurs at the flame tip and moves down as velocity increases.

- The position of transition from laminar to turbulent is called break point.

- Break point remains constant beyond certain velocity.

- The transition Reynolds number is different for different fuels.

Gaseous Jet Flame

- Gaseous single jet flame.

- Bunsen burner (with the vents closed).

- Two concentric streams of fuel and air.

Physical Description of a Jet Flame

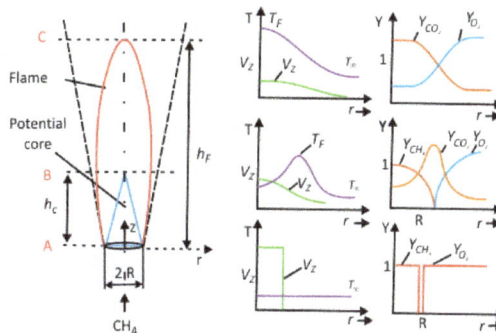

In a jet flame,

Fuel → Diffuses radially outward.

Oxidizer → Diffuses radially inward.

Mixing → Takes place at certain radial location and thus flame is established.

Flame Surface

- Location where fuel and oxidizer meet in stoichiometric proportion.

- As the reaction zone is quite thin, flame can be assumed as a thin surface.

Reaction Zone

Occurs in an annular region inside the flame surface.

Soot

Emission of soot in the fumes of a large diesel truck, without particle filters

Soot is a mass of impure carbon particles resulting from the incomplete combustion of hydrocarbons. It is more properly restricted to the product of the gas-phase combustion process but is commonly extended to include the residual pyrolysed fuel particles such as coal, cenospheres, charred wood, and petroleum coke that may become airborne during pyrolysis and that are more properly identified as cokes or chars.

Soot causes cancer and lung disease, and is the second-biggest human cause of global warming.

Sources

Soot as an airborne contaminant in the environment has many different sources, all of which are results of some form of pyrolysis. They include soot from coal burning, internal-combustion engines, power-plant boilers, hog-fuel boilers, ship boilers, central steam-heat boilers, waste incineration, local field burning, house fires, forest fires, fireplaces, and furnaces. These exterior sources also contribute to the indoor environment sources such as smoking of plant matter, cooking, oil lamps, candles, quartz/halogen bulbs with settled dust, fireplaces, exhaust emissions from vehicles, and defective furnaces. Soot in very low concentrations is capable of darkening surfaces or making particle agglomerates, such as those from ventilation systems, appear black. Soot is the primary cause of "ghosting", the discoloration of walls and ceilings or walls and flooring where they meet. It is generally responsible for the discoloration of the walls above baseboard electric heating units.

The formation of soot depends strongly on the fuel composition. The rank ordering of sooting tendency of fuel components is: naphthalenes → benzenes → aliphatics. However, the order of sooting tendencies of the aliphatics (alkanes, alkenes, and alkynes) varies dramatically depending on the flame type. The difference between the sooting tendencies of aliphatics and aromatics is thought to result mainly from the different routes of formation. Aliphatics appear to first form acetylene and polyacetylenes, which is a slow process; aromatics can form soot both by this route and also by a more direct pathway involving ring condensation or polymerization reactions building on the existing aromatic structure.

Description

The formation of soot is a complex process, an evolution of matter in which a number of molecules undergo many chemical and physical reactions within a few milliseconds. Soot is a powder-like form of amorphous carbon. Gas-phase soot contains polycyclic aromatic hydrocarbons (PAHs). The PAHs in soot are known mutagens and are classified as a "known human carcinogen" by the International Agency for Research on Cancer (IARC). Soot forms during incomplete combustion from precursor molecules such as acetylene. It consists of agglomerated nanoparticles with diameters between 6 and 30 nm. The soot particles can be mixed with metal oxides and with minerals and can be coated with sulfuric acid.

Soot Formation Mechanism

Many details of soot formation chemistry remain unanswered and controversial, but there have been a few agreements:

- Soot begins with some precursors or building blocks.

- Nucleation of heavy molecules occurs to form particles.

- Surface growth of a particle proceeds by adsorption of gas phase molecules.

- Coagulation happens via reactive particle–particle collisions.

- Oxidation of the molecules and soot particles reduces soot formation.

Hazards

Soot, particularly diesel exhaust pollution, accounts for over one quarter of the total hazardous pollution in the air.

Among these diesel emission components, particulate matter has been a serious concern for human health due to its direct and broad impact on the respiratory organs. In earlier times, health professionals associated PM10 (diameter < 10 μm) with chronic lung disease, lung cancer, influenza, asthma, and increased mortality rate. However, recent scientific studies suggest that these correlations be more closely linked with fine particles (PM2.5) and ultra-fine particles (PM0.1).

Long-termexposure to urban air pollution containing soot increases the risk of coronary artery disease.

Dieselexhaust (DE) gas is a major contributor to combustion-derived particulate-matter air pollution. In human experimental studies using an exposure chamber setup, DE has been linked to acutevasculardysfunction and increased thrombus formation. This serves as a plausible mechanistic link between the previously described association between particulate matter air pollution and increased cardiovascular morbidity and mortality.

Soot also tends to form in chimneys in domestic houses possessing one or more fireplaces. If a large deposit collects in one, it can ignite and create a chimney fire. Regular cleaning by a chimney sweep should eliminate the problem.

Soot Modeling

Soot mechanism is difficult to model mathematically because of the large number of primary components of diesel fuel, complex combustion mechanisms, and the heterogeneous interactions during soot formation. Soot models are broadly categorized into three subgroups: empirical (equations that are adjusted to match experimental soot profiles), semi-empirical (combined mathematical equations and some empirical models which used for particle number density and soot volume and mass fraction), and detailed theoretical mechanisms (covers detailed chemical kinetics and physical models in all phases) are usually available in the literature for soot models.

Empirical models use correlations of experimental data to predict trends in soot production. Empirical models are easy to implement and provide excellent correlations for a given set of operating conditions. However, empirical models cannot be used to investigate the underlying mechanisms of soot production. So, these models are not flexible enough to handle changes in operating conditions. They are only useful for testing previously established designed experiments under specific conditions.

Second, semi-empirical models solve rate equations that are calibrated using experimental data. Semi-empirical models reduce computational costs primarily by simplifying the chemistry in soot formation and oxidation. Semi-empirical models reduce the size of chemical mechanisms and use simpler molecules, such as acetylene as precursors. Detailed theoretical models use extensive chemical mechanisms containing hundreds of chemical reactions in order to predict concentrations of soot. Detailed theoretical soot models contain all the components present in the soot formation with a high level of detailed chemical and physical processes.

Such comprehensive models (detailed models) usually take high financial burden for programming and operating, and much computational time to produce a converged solution. On the other hand, empirical and semi-empirical models ignore some of the details in order to make complex model simple and to reduce the computational cost and time. Thanks to recent technological progress in computation, it becomes more feasible to use detailed theoretical models and obtain more realistic results. However, further advancement of comprehensive theoretical models must be preceded by the more detailed and accurate formation mechanisms.

On the other hand, models that are based on a phenomenological description have found wide use recently. Phenomenological soot models, which may be categorized as semi-empirical models, correlate empirically observed phenomena in a way that is consistent with the fundamental theory, but is not directly derived from the theory. Phenomenological models use sub-models developed to describe the different processes (or phenomena) observed during the combustion process. These sub-models can be empirically developed from observation or by using basic physical and chemical relations. Advantages of phenomenological models are that they are quite reliable and yet not so complicated. So, they are useful, especially when the accuracy of the model parameters is low. For example, the phenomenological models can predict the soot formation even when several operating conditions are changed in a system and the accuracy cannot be guaranteed. Examples of sub-models of phonological empirical models could be listed as spray model, lift-off model, heat release model, ignition delay model, etc.

Processes During Droplet Combustion

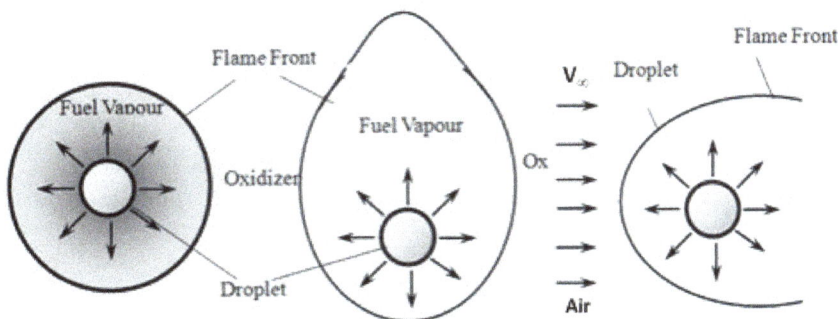

Factors affecting the shape of the flame front:

Condition under which combustion takes place

Zero gravity : Spherical flame front (No buoyancy)

Normal gravity : Elongated (Due to natural convection)

Forced convection condition: Fame aligned with flow

Energy required to vaporize the liquid fuel:

$$Qv = \Delta H_v + C_L \left(T_S - T_\infty \right)$$

Latent heat of vaporization   Sensible enthalpy

Liquid Fuel Combustion

Assumptions

1. Single droplet in quiescent atmosphere.

2. Droplet temperature is uniform.

3. Density of liquid fuel much higher than the gas phase.

4. Fuel is a single component with no solubility for gases.

5. Flow velocities are assumed to be low.

6. Single step irreversible reaction! Thin flame approximation.

7. Constant thermo-physical properties.

8. Unity Lewis number.

9. Radiation heat transfer is neglected.

10. No other phase is formed in the liquid fuel droplet.

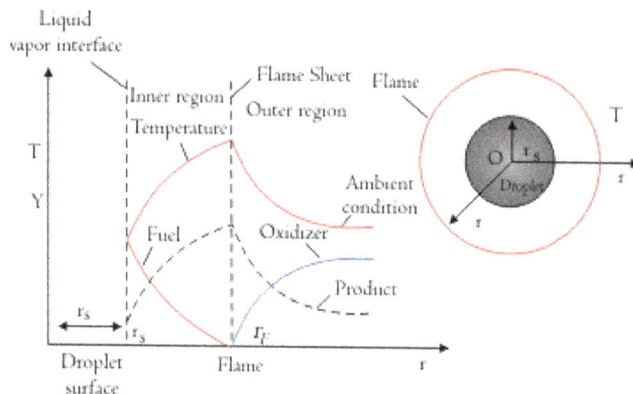

Fuel, oxidizer and product can be related to heat release rate as follows

$$-\dot{m}_F''' = \frac{-\dot{m}_F'''}{f} = \frac{-\dot{m}_p'''}{1+f} = \frac{\dot{q}'''}{f\Delta\hat{H}_C} \tag{1}$$

Rearranging,

$$\dot{q}''' + \Delta\hat{H}_C\dot{m}_F''' = 0 \tag{2}$$

We can rewrite fuel species conservation equation as,

$$\dot{m}'' r^2 \frac{dY_F}{dr} = \rho D \frac{d}{dr}\left(r^2 \frac{dY_F}{dr}\right) + \dot{m}_F''' r^2 \tag{3}$$

Multiply eq.3 by ΔH_C and add,

$$\dot{m}'' r^2 \frac{d(C_pT)}{dr} + \dot{m}'' r^2 \frac{d(Y_F\Delta\hat{H}_C)}{dr} = \rho\alpha\frac{d}{dr}\left(r^2\frac{d(C_pT)}{dr}\right) + \rho D\frac{d}{dr}\left(r^2\frac{d(Y_F\Delta\hat{H}_C)}{dr}\right) + r^2\left(\dot{q}''' + \dot{m}_F'''\Delta\hat{H}_C\right) \tag{4}$$

Here a is the thermal diffusivity

$$\left(k_g / \rho C_p\right); \alpha = D(Le = 1)$$

Using eq.2, eq.4 becomes,

$$\dot{m}'' r^2 \frac{d\left(C_pT + Y_F\Delta\hat{H}_C\right)}{dr} = \rho\alpha\frac{d}{dr}\left(r^2\frac{d\left(C_pT + Y_F\Delta\hat{H}_C\right)}{dr}\right) \tag{5}$$

Elimination of the non-linear term simplifies the analysis. This simplification is known as Schwab- Zeldovich Transformation.

Dividing eq.5 by , $Q_v + \Delta\hat{H}_C\left(Y_{FS} - 1\right)$,

$$\dot{m}'' r^2 \frac{db_{FT}}{dr} = \rho\alpha\frac{d}{dr}\left(r^2\frac{db_{FT}}{dr}\right) \tag{6}$$

$$b_{FT} = \frac{C_pT + Y_F\Delta\hat{H}_C}{Q_V + \Delta H_C\left(Y_{FS} - 1\right)}$$

Q_V – Heat input required for vaporization of droplet

Y_{FS} – Mass fraction of species at the surface of the droplet

$$Q_V + \Delta \hat{H}_V + C_L \left(Y_{Fs} - 1 \right)$$

Conserved variable for oxidizer

$$b_{ox,T} = \frac{C_p T + Y_{0x} \Delta \hat{H}_C}{Q_V + \Delta H_C f Y_{ox,s}} \qquad b_{F,ox} = \frac{Y_F + Y_{ox,f}}{\left(Y_{Fs} - 1 \right) + Y_{0x,s} f}$$

General format of all the equations

$$\dot{m}'' r^2 \frac{db}{dr} = \rho \alpha \frac{d}{dr} \left(r^2 \frac{db}{dr} \right) \tag{7}$$

Boundary conditions

$$r = r_s \quad \dot{m}_F'' = \rho \alpha \left(\frac{db}{dr} \right)_{r=rs} \quad ; \qquad r \to \infty; b = b_\infty$$

Integrating eq. 7 twice and applying the boundary conditions,

$$\frac{\dot{m}_F'' r_s^2}{\rho \alpha} \frac{1}{r} = \ln \left(\frac{b_\infty - b_s + 1}{b - b_s + 1} \right) \tag{8}$$

At $r = r_s; b = b_\infty$

By applying boundary condition to Eq. 8, We can have,

$$\dot{m}_F''' = \frac{\rho \alpha}{rs} \ln \left(1 + B \right)$$

The transfer number, B is given by

$$B_{F,T} = \frac{C_P \left(T_\infty - T_s \right) + \Delta \hat{H}_C \left(Y_{F\infty} - Y_{F,s} \right)}{Q_v + \Delta \hat{H}_c \left(Y_{Fs} - 1 \right)}$$

$$B_{ox,T} = \frac{C_P \left(T_\infty - T_s \right) + \Delta \hat{H}_C \left(Y_{ox,\infty} - Y_{ox,s} \right)}{Q_v + f \Delta \hat{H}_c \left(Y_{ox,s} \right)}$$

$$B_{F,ox} = \frac{\left(Y_{F,\infty} - Y_{F,s} \right) + \left(Y_{ox,s} - Y_{ox,\infty} \right) f}{\left(Y_{F,s} - 1 \right) + f \left(Y_{ox,s} \right)}$$

Values of transfer number, B for some typical fuel:			
Combustion in air	**B**	**Combustion in air**	**B**
ISO-Octane	6.41	Kerosene	3.4
Benzene	5.97	Gas oil	2.5
n-Heptane	5.82	Light fuel oil	2.0
Avation gasoline	5.5	Heavy fuel oil	1.7
Automobile gasoline	5.3		

References

- Glassman, Irvin; Yetter, Richard A. (2008). "6. Diffusion Flames". Combustion. Burlington: Academic Press. ISBN 978-0-12-088573-2

- Niessner, R. (2014), The Many Faces of Soot: Characterization of Soot Nanoparticles Produced by Engines. Angew. Chem. Int. Ed., 53: 12366–12379. doi:10.1002/anie.201402812

- Juliet Eilperin (2013-11-26). "Black carbon ranks as second-biggest human cause of global warming". The Washington Post. Retrieved 2013-12-04

- Omidvarborna; et al. "Recent studies on soot modeling for diesel combustion". Renewable and Sustainable Energy Reviews. 48: 635–647. doi:10.1016/j.rser.2015.04.019

- Omidvarborna; et al. "Characterization of particulate matter emitted from transit buses fueled with B20 in idle modes". Journal of Environmental Chemical Engineering. 2 (4): 2335–2342. doi:10.1016/j.jece.2014.09.020

Permissions

All chapters in this book are published with permission under the Creative Commons Attribution Share Alike License or equivalent. Every chapter published in this book has been scrutinized by our experts. Their significance has been extensively debated. The topics covered herein carry significant information for a comprehensive understanding. They may even be implemented as practical applications or may be referred to as a beginning point for further studies.

We would like to thank the editorial team for lending their expertise to make the book truly unique. They have played a crucial role in the development of this book. Without their invaluable contributions this book wouldn't have been possible. They have made vital efforts to compile up to date information on the varied aspects of this subject to make this book a valuable addition to the collection of many professionals and students.

This book was conceptualized with the vision of imparting up-to-date and integrated information in this field. To ensure the same, a matchless editorial board was set up. Every individual on the board went through rigorous rounds of assessment to prove their worth. After which they invested a large part of their time researching and compiling the most relevant data for our readers.

The editorial board has been involved in producing this book since its inception. They have spent rigorous hours researching and exploring the diverse topics which have resulted in the successful publishing of this book. They have passed on their knowledge of decades through this book. To expedite this challenging task, the publisher supported the team at every step. A small team of assistant editors was also appointed to further simplify the editing procedure and attain best results for the readers.

Apart from the editorial board, the designing team has also invested a significant amount of their time in understanding the subject and creating the most relevant covers. They scrutinized every image to scout for the most suitable representation of the subject and create an appropriate cover for the book.

The publishing team has been an ardent support to the editorial, designing and production team. Their endless efforts to recruit the best for this project, has resulted in the accomplishment of this book. They are a veteran in the field of academics and their pool of knowledge is as vast as their experience in printing. Their expertise and guidance has proved useful at every step. Their uncompromising quality standards have made this book an exceptional effort. Their encouragement from time to time has been an inspiration for everyone.

The publisher and the editorial board hope that this book will prove to be a valuable piece of knowledge for students, practitioners and scholars across the globe.

Index

www.ingramcontent.com/pod-product-compliance
Lightning Source LLC
Chambersburg PA
CBHW061959190326
41458CB00009B/2912